Fractal Analysis - Selected Examples

Edited by Robert Koprowski

Published in London, United Kingdom

IntechOpen

Supporting open minds since 2005

Fractal Analysis - Selected Examples
http://dx.doi.org/10.5772/intechopen.87695
Edited by Robert Koprowski

Contributors
Haijun Luan, Dzhennet Magomedova, Oleg Sheluhin, Ricardo Fernández, Gaspar Gonzalez-Doncel, Gerardo Garcés, Rafik Menassel, Sanjeev Kimothi, Narendra Singh, Asha Thapliyal, Mohamed Abdelsabour Fahmy

First published in London, United Kingdom, 2020 by IntechOpen
IntechOpen is the global imprint of INTECHOPEN LIMITED, registered in England and Wales, registration number: 11086078, 7th floor, 10 Lower Thames Street, London, EC3R 6AF, United Kingdom
Printed in Croatia

British Library Cataloguing-in-Publication Data
A catalogue record for this book is available from the British Library

Additional hard and PDF copies can be obtained from orders@intechopen.com

Fractal Analysis - Selected Examples
Edited by Robert Koprowski
p. cm.
Print ISBN 978-1-83962-482-7
Online ISBN 978-1-83962-483-4
eBook (PDF) ISBN 978-1-83962-484-1

For EU product safety concerns:
IN TECH d.o.o., Prolaz Marije Krucifikse Kozulić 3, 51000 Rijeka, Croatia, info@intechopen.com or visit our website at intechopen.com.

Meet the editor

Robert Koprowski, MD (1997), PhD (2003), Habilitation (2015), is an employee of the University of Silesia in Poland, Institute of Computer Science, Department of Biomedical Computer Systems. For 20 years he has studied the analysis and processing of biomedical images with particular emphasis on the full automation of measurement for a large inter-individual variability of patients. He is the author of dozens of papers with the impact factor (IF) and more than a hundred other papers, as well as the author or co-author of six books. Additionally, he is the author of several national and international patents in the field of biomedical devices and imaging. Since 2011, he has been a reviewer of grants and projects (including EU projects) in the field of biomedical engineering.

Contents

Preface XI

Chapter 1 1
Optimization of Fractal Image Compression
by Rafik Menassel

Chapter 2 21
Fractal Analysis for Time Series Datasets: A Case Study of Groundwater Quality
by Sanjeev Kimothi, Asha Thapliyal and Narendra Singh

Chapter 3 39
Establishing the Downscaling and Spatiotemporal Scale Conversion Models of NDVI Based on Fractal Methodology
by Haijun Luan

Chapter 4 55
A New BEM for Modeling and Optimization of 3T Fractional Nonlinear Generalized Magneto-Thermoelastic Multi-Material ISMFGA Structures Subjected to Moving Heat Source
by Mohamed Abdelsabour Fahmy

Chapter 5 85
Using Algebraic Fractals in Steganography
by Oleg Sheluhin and Dzhennet Magomedova

Chapter 6 103
Fractal Analysis of Strain-Induced Microstructures in Metals
by Ricardo Fernández, Gaspar González-Doncel and Gerardo Garcés

Preface

Fractal analysis is becoming more and more common in all walks of life. This includes biomedical engineering, steganography and art. Writing one book on all these topics is a very difficult task. For this reason, this book covers only selected topics. Interested readers will find in this book the topics of image compression, groundwater quality, establishing the downscaling and spatio-temporal scale conversion models of NDVI, modelling and optimization of 3T fractional nonlinear generalized magneto-thermoelastic multi-material, algebraic fractals in steganography, strain induced microstructures in metals and much more.

For example, one of the chapters covers the application of fractals to steganography. As is well known, steganography includes the technology of hiding watermarks in multimedia files. This technology is important in the field of copyright protection, secret communication and the form of information exchange. Contemporary image analysis and processing methods allow the determination of the presence of embedded watermarks in a stegoimage on the basis of the original image and its statistical features. In this chapter, the authors propose the use of algebraic fractals for steganographic embedding of watermarks in colour images. The authors propose the use of algebraic fractals for an image serving as a secret key, which, according to the authors, allows for more resistant deposition to computer attacks, including compression.

In another chapter, the author describes the use of fractals to analyze the terrain surface using remote sensing. The author created the downscaling models using fractal theory for the parameters of the earth's surface in quantitative remote sensing. Additionally, it is still a hotspot for creating space-time conversion model parameters of the land surface in quantitative remote sensing in the future and marked in this connection, the author developed a multifractal scaling methodology and its availability is proposed.

This book brings together a small collection of chapters covering specific aspects of the fractal field. I would like to thank the publishers (IntechOpen) for their help in preparing this book and the authors of the accepted manuscripts for their work and patience. For interested readers, I would like to suggest that they pay attention to the annual conferences on fractals and their use in various areas of life.

The book will definitely be of interest to scientists dealing with fractal analysis, as well as biomedical engineers or IT engineers. I encourage you to view individual chapters.

Robert Koprowski
Faculty of Science and Technology,
University of Silesia,
Sosnowiec, Poland

Chapter 1

Optimization of Fractal Image Compression

Rafik Menassel

Abstract

Fractal encoding is a promising method of image compression. It is built on the basis of the forms found in the image and the generation of repetitive blocks based on mathematical translations. The technique seems to be moved theoretically and practically, but it requires enormous programming time due to the excessive resources required when compressing large volumes of data. On the other hand, metaheuristics represent all of the methods used to solve difficult optimization problems with less consumption of resources. They are marked by their rapid convergence and their lessening in research difficulties. In this chapter, we have tried to apply a new experience around the performance of organic metaheuristics inspired by nature, which are, respectively, the wolf pack algorithm (WPA) and the bat-inspired algorithm (BIA), as bioinspired techniques to optimize the fractal image compression (FIC). Experiments show the enhancement of diverse characteristics (coding time, compression rate (CR), peak signal-to-noise ratio (PSNR), and mean square error (MSE)). In addition, an assessment of the proposed approaches via many other approaches highlights this improvement.

Keywords: fractal image compression (FIC), metaheuristics, wolf pack algorithm (WPA), bat-inspired algorithm (BIA), image quality

1. Introduction

Nowadays, a significant size of information is managed and transmitted, and mainly, images have involved prodigious status, particularly in recognition field. Thus, it is important to decrease the size of the data via compression algorithms which can allow their storage and their transmission while using limited resources. Compression is employed to overcome this problem and keep more files. Mainly multimedia files need more storage space than other types of files. Images represent the largest part of the most used multimedia files in almost all fields. Unlike other types of files, a huge amount of image data requires more resources for storage and transmission on computer networks, and compression is therefore presented as an inevitable tool with the aim of more maneuverability of this data. Today, several compression formats exist while presenting their limits (degradations, size, duration, etc.) on somewhat particular images (text images and background areas).

To overcome this difficulty, scientists are constantly developing new techniques to compress images in order to find a perfectionist compression method that can largely conserve storage space and preserve the quality of the source file.

The compression methods that exist tend to introduce the theory of fractals, which appears to be a strong instrument for boosting image quality and reducing the resources required. Nowadays, the image files intricated in several specific programs are characterized by massive data, pixel correlation, and similarity. Under such conditions, the old compression methods appear to be unsuitable for this mission because of their need for significant encoding time. On the other hand, the recently developed fractal image compression techniques offer better compression qualities [1–3]. These methods are built on the principle that shapes (fractals) can better represent usual scenes than traditional geometric forms. A number of fractal image compression techniques are presented in a lot of works; they include an image coding agreeing to more security and less degradation (Jeng et al. [4], Li [5], and Han [6]). These methods have revealed an important in information, reduced statistical characteristics of chaotic patterns, and weakness in statistical cryptanalysis. In addition, metaheuristics are likewise employed to optimize compression, like genetic algorithms [7], ant colonies [8], and optimization by particle swarms [9]. These methods have ability to create a partition constructed on a region which improves the compression ratio and maintains better the decompressed image quality.

In this chapter, we try for the first time to apply natural metaheuristics on fractal compression, by suggesting new methods which associate, both the bat-inspired algorithm (BIA) and the wolf pack algorithm (WPA) with fractal image compression (FIC) to speed up encoding and optimize both file size and image quality. The main objective of using these algorithms is its property of search for global solution and its capacity to generate very satisfactory results quickly and for less means compared to similar techniques. The algorithms also use fewer parameters and without initial approximation of the unidentified parameters. This document is organized as follows; Section 1 presents an introduction in the study context. Next, a summary of the fractal image compression is detailed in Section 2. Section 3 summarizes the natural metaheuristics and presents those used in this work, namely, the WPA and the BIA. In Section 4, we give a review of some related works. After that, we present our proposed methods in Sections 5 and 6. Experiments are explained in Section 7. Finally, the conclusion appears in Section 8.

2. Fractal encoding

Fractal compression is a new method of irreversible image compression [10]. It was adapted by *Hutchinson* [11] and *Barnsley and Demko* [12]. It searches for self-similarities among the diverse image blocks [13] and only keeps the parameters of the contractual transformation in place of the pixels of the image. Like this, we can build an estimate as close as possible to the source image by detecting the redundancy of forms at several scales and try to eliminate these redundancies in the original image so that the result is precise enough to be accepted.

The FIC is founded on an iterated function system *f*, a limited group of contractions defined on a metrical space R_n by the relation:

$$f_i : R_n \rightarrow R_n; i < N \tag{1}$$

This contraction can be in several shapes depending on technical constraints. It can be done several points of the original image and carry them nearer to the compressed one. This reflection is named "affine transformation"; then, each sub-block of the original image will be submitted either a rotation at an angle, a scale, or a translation (transformed using eight isometries) according to Equation 2:

$$w(x) = T(x) + b \tag{2}$$

where T is a linear transformation, $R_n \rightarrow R_n$ is a vector, and $b \in R_n$ is a vector. Practically, the general principle of fractal compression is to try to find the finest matching domain blocks for each range block in order to minimize the distance metric. The code that follows illustrates this idea well:

- Enter the original image
- Create a partitioning of Range Blocks R
- Create a partitioning of Domain figures
- ***For*** all figures Destination Make
    - ***For*** all figures Sources Make
- ***For*** all the defined transformations Make
    - Apply the transformation to the Range Blocks
    - Adjust the average of the pixel colors
    - Apply the reduction from the Range Blocks to the Domain Blocks
    - Calculate the error between the result and the Domain Blocks
    - ***If*** the error is minimal for the destination figure ***Then***
- Save the modifications made
    - ***End if***
- ***End For***
- Write the saved values in the output file
    - ***End For***
- ***End For***

This process is realized following the relation:

$$B_i = v(D_i) \tag{3}$$

where v () is the function of the contraction which aims to modify D_i. Thereafter, the nearest block B_i is sought for all R_i blocks by calculating the error between B_i and R_i, and we can use, for example, the Hausdorff distance defined by Equation 4:

$$H(B_i, R_i) = \max\left(d(B_i, R_i), d(R_i, B_i)\right) \tag{4}$$

where d(B, R) = max b ϵ A min r ϵ B ||b-r||.
Or using the Euclidean distance described by the equation:

$$d^2(R, B) = \sum^{n} (r_i, b_i)^2 \quad (5)$$

where n represents the pixel's number in R_i and B_i blocks. d should be as minimum as possible for blocks that look alike.

Fractal decompression consists of the reconstruction of R_i from the blocks B_i which appear identical the most by practicing the contraction used in compression.

In recent years, several studies have concentrated on the development of accelerated decryption procedures [14–16] with the aim of preserving image quality. Their principle is to choose a random image as the original one and execute an affine transformation like defined in Equations (6) and (7), founded on the fractal ciphers obtained by itself. This act is repeated recursively till the reconstructed image were satisfactory:

$$R_i = S_i.D + O_i.I \quad (6)$$

$$S = U\ R_i \quad (7)$$

where I is a contractive or isometric spatial transformation, D is a domain block, R is a range block, and ***S*** is the reconstructed image.

In fact, FIC is becoming among the most promising methods for image compression for its significant compression ratio (CR) and preservation of quality. Its beginnings date from the 1990s when Jacquin [17, 18] introduced the first method of image compression; its principle is partitioning the image into two tiling blocks: the range and domain blocks.

The domain blocks are double the size of the range blocks and overlap such that a new domain block starts at each pixel. The main idea of this compression is to find the nearest domain block in concordance with each range block, to determine the right contractual transformation, and to store all these parameters. This principle was exciting; however it remained limited to local applications because it consumes a lot of CPU time. Since then, researchers have constantly presented new techniques to reduce the compression time; Thomas and Deravi [19] link range blocks and brand them more adaptive with image content by using the region-growing method. Cardinal [20] presented an alike principle; it is founded on a geometrical partition of the grayscale image block feature space. The experimental evaluations with earlier published methods illustrate an important enhancement in encoding time with practically better quality. *He et al.* [21] have used the normalized block with the aim to evade the extreme search in corresponding block. Chong and Pi [22] proposed a new adaptive search method to decrease the calculation complexity of fractal encoding to discard a big number of unqualified domain blocks so as to speed up FIC.

Other studies have been presented on new aspects to improve the way of research like the encoding via the Fourier transform [23], special image features [24], and discrete cosine transform inner product [25]. The majority of existing methods rely on a corresponding error threshold to limit the search. Lately, Lin and Wu [26] have defined another way of search built on image block edge property, which proves suitable results. Furthermore, many research articles have been published over the past decade; they increased the quality of the compressed image through the use of metaheuristics without resorting to more resources in the coding process.

3. Natural heuristics

Heuristics refer to the set of techniques that can solve several problems by maximizing gains and decreasing the resource's consumption; however, the optimal solution cannot be certain if, in the investigation space, there is an intersection between the local and global solution [27]. Natural heuristics represent a large family of heuristics which are inspired from communal conduct of animals existing in societies like assemblages of birds, ant colonies, or grouping of fish. They are founded on the principle of populations of entities who cooperate and develop rendering to reciprocal precepts. These techniques allow the invention of procedures which can resolve hard problems by dividing control. These approaches form a famous prototype which is effectively employed as a prodigious tool for resolving difficult problems [28] with less resource consumption. Several researches [29, 30] illustrate that these systems have an effective potential to manage various situations and could be adapted to bring solutions to diversified optimization problems.

3.1 The wolf pack algorithm (WPA)

The wolf pack algorithm (WPA) [31] belongs to the family of bioinspired techniques which can be used to estimate resolutions for numerous optimization problems. WPA is a metaheuristic built on the population invoked by the social hunting behavior of wolves. It basically involves hunting wolves, tracking down prey, and capturing it under the orders of a leader wolf. The wolf pack includes the strongest and most intelligent wolf chef. He is responsible for controlling the pack. Its decisions are always based on the surrounding environment: prey, pack wolves, and other hunters. The pack is divided into two families of wolves: scoot and ferocious.

The scoot wolves move autonomously in the milieu and adjust its way according to the concentration of the odor of the prey. When a prey is found, scoot wolves cry and transmit info by sound to the leading wolf who guesses the distance to reach this prey; it calls the furious wolves and quickly displaces towards the cry. The prey is then caught and is shared conferring to the nature of each wolf: from the sturdiest to the feeblest. Subsequently, feeble wolves could die from absence of nutrition. In this manner the pack ensures a certain dynamic and robustness at all times.

WPA is performed as follows:

In a search environment $\mathbb{R}^n$, any wolf $\boldsymbol{i}$ denotes an elementary solution to the problem, at a location x_i.

Initially, wolves are distributed chaotically in the environment.

At any instant t, the wolf i passes from the location x^t_i to the location x^{t+1}_i. The choice of the following location is updated by rendering the following equation:

$$x_i^{(t+1)} = x_i^{\,t} + \lambda\,|x_g^{\,t} - x_i^{\,t}\,| \tag{8}$$

where $\boldsymbol{\lambda}$ represents a vector randomly distributed in the interval $[-1,1]$ and $\boldsymbol{x_g}$ designates the location of the chief wolf. After a static sum of repetitions, which corresponds to a research stage, the wolf of the finest result gets converted to a leader one; feeble wolves (bad solutions) will be wiped out and substituted with a novel group of wolves in an arbitrary manner.

3.2 The bat-inspired algorithm

The bat-inspired algorithm (BIA) [32] belongs to the family of metaheuristics inspired by nature, introduced by Yang and founded on the echolocation comportment of bats. Bats have a system identical to radar except that radars use electromagnetic waves while bats use ultrasonic waves (of frequency inaudible to humans). Bats move and hunt with high-performance sonar. By another way, bats are distinguished by an extraordinary steering mechanism allowing them to differentiate between an obstacle and a prey, which allows them to hunt even all. The impulses produced by bats can be linked almost to the hunting plans of bats. Often, the pulses are between 25 and 150 kHz at a static frequency; they only persist for 8–10 ms. Bats generate between 10 and 20 ultrasonic sound eruptions every second, each of them stays between 5 and 20 ms. However, when bats seek their prey and feel so close, they can increase the rate of emission of sound eruptions up to 200/s. This proves the extraordinary ability of bats to process signals. Assuming that the speed of sound in air is v = 340 m/s, the wavelength λ of the ultrasonic sound therefore manifests with a continuous frequency *f*:

$$\lambda = v/f \tag{9}$$

The wavelengths vary between 2 and 14 mm and are equivalent to the size of the bat's prey, for a representative frequency between 25 and 150 kHz. The pulses produced by bats can spread an imposing sound intensity of 110 dB, but quite favorably, these pulses remain in the ultrasonic domain. The intensity of the pulse can take different stages, such as very strong when bats are chasing and weak at a quiet sound when they mark their prey. These short pulsations usually have a wandering range of a few meters which depends on the frequency.

In reality, bats combine all of their senses to effectively detect prey and navigate more easily. Here, we are only interested in echolocation and the behaviors that accompany it. To create new optimization techniques, the echolocation of bats can be transformed into an optimized objective function.

In a search space R_i, the bats fly randomly using the speed Vi in location (solution) X_i using velocity V_i. They produce pulses at a static wavelength λ with a variable frequency f and an intensity A (differs from a big positive A_0 to a smallest constant value A_{min}) to hunt for prey. When the bats choice the finest results, they choose a local result from the best selected ones.

4. Introducing bioinspired approaches in the FIC

In 2005, Dervis Karaboga proposed a new iterative optimization method based on artificial bee colonies (ABC). This technique is based on three different classes of bees, (a) bee used, (b) spectator bee, and (c) scout bee. The spectator bees waiting in the store obtain data concerning the sources of nectar revealed earlier from the employees. Then they choose a usable nutrition source built on the received information. Scout bees arbitrarily search for nutriment in the area for [33].

In 2006, Cristian Martinez presented an enhanced image compression using the ant colony technique. The basic idea is that ants always seek and find the shortest path from nest to food source using the pheromone. For fractal compression, the pheromone is positioned on the range block i and the domain block j. The pheromone matrix is rectangular (not symmetrical) where the lines designate range blocks (image blocks) and the columns indicate domain blocks (blocks to transform). Then, the ants build routes by choosing a block of domain j for each block of range i. the solution will be found on the basis of updating the pheromone and heuristic information [34]. The result proposes similar image quality to that obtained with a deterministic way while minimizing the calculation time by 34%.

In 2009, many of studies were focalized on FIC: Chakrapani and Soundara Rajan [35] have created a new fractal image compression founded on a genetic algorithm in the intention of optimizing the encoding time for an acceptable image quality. The results give improved performance over exhaustive search.

In the work of Xing-yuan, Fan-ping, and Shu-guo [36], a spatial correlation hybrid genetic algorithm that uses the features of the fractal and divided iterative function system is proposed. It consists of two stages. The first uses spatial correlation in the images for the range and the domain blocks in order to exploit the local optima. The second one is based on a genetic-simulated annealing algorithm (SAGA) to find the global optima if the local optima is not satisfied. In order to escape early convergence, the algorithm approves that the dyadic mutation operator takes place in place of the traditional operator.

In 2010, Chakrapani et al. have enhanced the fractal image compression using particle swarm optimization (PSO) technique [37]. PSO is used to speed up the search of the nearest finest match block for a definite block to be encoded. This method illustrates that the recovered image quality can be conserved when in comparison with the full-search FIC.

In 2016, Shaimaa S. Al-Bundi et al. [38] use an upgraded genetic algorithm to enhance the exploration space in the target image by good estimation to the global optimum in an only execution.

In 2017, Al-Saidi N.M.G et al. [39] optimize the fractal image compression by introducing the harmony search algorithm. This strategy searches for the best solution through singing a song; this proposed technique offers splendid performances in terms of image quality, reduced computation time, and storage space when compared to other methods.

In 2018, we [40] used the wolf pack algorithm to improve the FIC; the idea is to take the entire image for the search space where this space is divided into blocks; scooters wolves roam the environment to find other smaller and similar blocks. They examine the entire space and select the blocks with the best physical shape. By this method, the encoding time was considerably reduced, and we also obtained a better compression rate.

In 2019, we have [41] chosen to improve the FIC by using the bat-inspired algorithm. Our tow proposed methods are detailed and well explained in Sections 5 and 6, respectively.

5. Wolf pack algorithm to optimize fractal image compression

We assume an image of m x n pixels as exploration space, represented by an array P where each pixel is considered as a cell and on a byte (gray pixel). The resulting image C of $m/2 \times n/2$ pixels is reached by following the steps:

- Divide the entire image into tiny nonoverlapping r_i blocks of size $s \times s$ (with $s << m$). More simply, we will proceed with blocks of square size of b x b; this partition called range block will be represented by $R_N = \{r_1, r_2, \dots, r_N\}$.

- For all the blocks r_i, the scooting wolves roam the space in order to find a di of size $2b \times 2b$ similar with r_i while respecting the parameters mentioned above. A fitness value f (di) will be assigned for each block d_i according to Eq. (6). The block d_i is taken for prey.

- After the hunting wolves have inspected the entire space and for all r_i blocks, d_i blocks with the best physical form are selected. It will be mapped according to Eqs. (4) and (5).

If no improvement is made to the wolf chef solution, the process will be stopped after a fixed number of iterations.

The adapted WPA algorithm for FIC is showed as:

Algorithm 1. FIC-WPA

```
Initialization:
Generate r_i, (i = 1 ... N)
For each r_i, f(r_i) ← 0, (i = 1 ... N)
    While not (stopping criteria)
    it←0
        While not (Iter-scoot < It)
        Pick random numbers: λ ϵ [−1,1]
            For each r_i do
                x_i ← x_i + λ|x_g − x_i|
                If (f(bx_i) < f(g)) g ← i
                Endif
            Endfor
        ri ← v(di)
        Update It
        EndWhile
    EndWhile
End For.
```

---

6. BAT-inspired algorithm introduced in FIC

The similar idea employed in the wolf pack algorithm and the BAT algorithm for FIC is completed through the succeeding phases:

1. The whole image is scouted randomly by bats with the use of loudness L and frequency F.

2. Each block is compared to all of its neighboring blocks by bats for its degree of homogeneity as a function of volume and frequency. If they meet a criterion (color_level_block - neighboring color level ≤ frequency), we create a domain block of size L * L whose value is only the mean of the domain block.

3. When the bats roam the whole picture. The iteration will be stopped.

4. After decomposing the image into domain blocks, the bats' position themselves at xi, and the size of the blksz block will be saved in a sparse S.

5. By eliminating the solution with the smaller block, we try to find the best solution in this step.

6. In order to calculate the compression ratio, we will use Huffman encoding to store all information (locations, block sizes, and values).

7. Finally, and to reconstruct the image, we will use Huffman decoding to regenerate the image data of the compressed image.

Algorithm 2. Proposed FIC-BAT

```
Initialization: Generate bats (Number_bats = 1..N) //
Begin
Loudness L;
Frequency F;
While not (stopping criteria)
   For each bat
       If similarity (distance) = 1
                  Create domain block;
                  Store location in vectors I,J;
                  Store block sizes in vector blksz;
       Else
                  Store location in vectors I,J;
                  Store block sizes in vector blksz;
       End-if
   End For
End-while
Search for best solutions;
Store the locations and block sizes in a sparse S;
End.
```

7. Experimentations

7.1 Fractal image compression with WPA

The resolution presents an aspect that should not be ignored when trying to test the efficiency of an approach. For our circumstance, we will take into consideration the three test images (Lena, Barbara, and cameraman) with different resolutions (16 * 16, 32 * 32, 64 * 64, 128 * 128, 256 * 256) so that we can see the impact of this factor on the quality quantities (compression ratio, compression time, EQM, PSNR) (**Figures 1** and **2**).

From the test images with a reconstructed resolution of 256 * 256, it can be seen that the quality of the images is acceptable to a very good degree. And to be more exact, we must refer to calculable measures.

The table below presents the results obtained by applying our approach to the above images with different resolutions:

(a) (b) (c)

Figure 1.
Tested images (before compression). (a) Cameraman, (b) Lena, and (c) Barbara.

(a) (b) (c)

Figure 2.
Decompressed tested images (after WPA compression). (a) Cameraman, (b) Lena, and (c) Barbara.

The results obtained (detailed in **Table 1**) illustrate firstly the agreement between image resolution and encoding time, which demonstrates that our proposed method is significantly responsive to image resolution; it is also clear that the image quality is in contrast linked to it (image quality degrades quickly as resolution increases). This is clearly shown by the amplitudes of the PSNR and the MSE.

The compression ratio remains relative to the resolution where we indicate that our method interferes in this measurement. Wolves appear when the quantity of blocks is very huge (higher resolution) by contributing a higher compression rate than that existing in lesser resolutions.

Decompression time stays optimized and is similar for almost all methods. This is due to the decompression process which is not complex compared to the compression process.

We will now focus on comparing our proposed method with some other techniques. The following table (**Table 2**) shows the notable alteration between our method and the others:

Tested image	Resolution	Compression time (s)	Decompression time (s)	Compression ratio	MSE	PSNR (dB)
Cameraman	16*16	0.184	0.818	1.098	2.617	37.999
	32*32	0.446	0.802	1.174	4.079	36.173
	64*64	2.451	0.816	1.420	6.853	31.509
	128*128	41.284	0.810	1.653	8.931	30.096
	256*256	774.438	0.897	1.741	9.152	31.605
Lena	16*16	0.185	0.832	1.095	7.730	33.616
	32*32	0.507	0.801	1.125	5.861	34.451
	64*64	2.290	0.791	1.287	8.401	33.550
	128*128	34.257	0.821	1.480	9.547	32.641
	256*256	668.810	0.850	1.596	9.282	33.305
Barbara	16*16	0.225	0.812	1.019	1.352	38.070
	32*32	0.607	0.763	1.055	4.064	36.067
	64*64	1.868	0.781	1.152	6.259	35.060
	128*128	22.713	0.793	1.310	11.093	32.348
	256*256	509.406	0.861	1.447	11.115	33.288

Table 1.
Variation in image resolution.

Images de test	Methods	PSNR (dB)	Time (s)	Ratio (CR)
Lena 128*128	FIC-WPA	32.641	34.257	1.480
	Suman K. Mitra et al.'s works [7]	30.22	/	1.059
	Vishvas V. Kalunge et al.'s works [42]	/	67	/
Lena 256*256	FIC-WPA	33.305	668.810	1.596
	Y. Chakrapani et al.'s works [37]	26.22	2370	1.3
	Exhaustive search	32.69	8400	1.3
	DWSR [31]	25.8212	56.4247	1.56355
	PSO-RCQP [43]	27.089	6.453	1.6392
Cameraman 256*256	FIC-WPA	31.605	774.438	1.741
	PSO-RCQP [43]	26.686	268	1.8212
Barbara 128*128	FIC-WPA	32.348	22.713	1.310
	Vishvas V. Kalunge et al.'s works [42]	/	66	/
Barbara 256*256	FIC-WPA	33.288	509.406	1.447
	Y. Chakrapani et al works [37]	28.34	2500	1.2
	Exhaustive search	32.84	8400	1.2

Table 2.
Comparison of the FIC-WPA with other techniques.

7.2 Using bat-inspired algorithm to enhance fractal image compression

Bat's number: In **Table 3**, we will examine the images of the cameraman and Lena in order to extract the adequate number of bats that will be used in our approach.

Loudness: Table 4 presents some tests carried out on an additional stress which is loudness, with the objective of taking into consideration the good value for the test.

Frequency: The last test table (**Table 5**) is made to pick up the best frequencies that will be used in our algorithm.

Images	Number of bats	Compression time	Decompression time	Compression ratio	PSNR	MSE
Cameraman	2	0.488	0.705	1.385	31.608	10.088
	4	0.459	0.707	1.366	30.934	8.563
	8	0.452	0.729	1.372	31.216	9.417
	16	0.451	0.749	1.349	30.934	8.879
	32	0.509	0.843	1.355	29.827	10.045
	64	0.472	0.919	1.365	30.412	10.405
	128	0.478	0.749	1.348	31.895	9.663
	256	0.475	0.883	1.335	30.989	8.899
	512	0.457	0.750	1.392	29.997	9.870
Lena	2	0.518	0.739	1.303	30.629	14.440
	4	0.548	0.727	1.315	30.083	15.612
	8	0.505	0.716	1.311	30.071	14.929

Images	Number of bats	Compression time	Decompression time	Compression ratio	PSNR	MSE
	16	0.514	0.713	1.306	29.984	15.348
	32	0.563	0.784	1.299	30.452	15.037
	64	0.562	0.779	1.298	31.228	13.887
	128	0.512	0.715	1.306	30.669	15.603
	256	0.518	0.709	1.320	30.389	16.118
	512	0.520	0.702	1.321	29.856	14.710

Table 3.
Testing the number of bats.

As we can see from the previous tables, the best choices are given by 8 for the number of bats, 8 for the intensity, and 30 for the frequency, so the treatment parameters are as follows:

The number of iterations is in the interval of [10,100].
The number of bats is fixed at 8.
Loudness 8, Frequency 30.

We experienced the proposed approach on five images through diverse resolutions. **Table 6** illustrates certain quality measurement liable on the resolutions.

Image	Loudness	Compression time	Decompression time	Compression ratio	PSNR	MSE
Cameraman	2	0.453	0.720	1.267	33.788	6.531
	3	0.452	0.729	1.372	31.216	9.417
	4	0.458	0.721	1.376	33.022	7.807
	5	0.469	0.702	1.355	30.220	9.375
	6	0.438	0.727	1.359	30.119	9.027
	7	0.460	0.723	1.348	28.971	9.130
	8	0.442	0.727	1.279	34.284	5.600
	9	0.455	0.736	1.220	30.813	6.962
	10	0.468	0.757	1.205	33.859	5.012
	11	0.472	0.755	1.187	33.548	4.362
Lena	2	0.530	0.735	1.248	31.781	14.848
	3	0.505	0.716	1.311	30.071	14.929
	4	0.537	0.726	1.324	29.230	16.853
	5	0.586	0.751	1.299	30.536	14.382
	6	0.519	0.713	1.244	30.812	13.766
	7	0.516	0.715	1.223	31.290	12.190
	8	0.503	0.756	1.228	31.032	11.917
	9	0.495	0.740	1.204	32.213	9.056
	10	0.492	0.761	1.194	31.873	8.372
	11	0.488	0.737	1.192	32.050	9.695

Table 4.
Testing the values of loudness.

Image	Frequency	Compression time	Decompression Time	Compression ratio	PSNR	MSE
Cameraman	20	0.490	0.714	1.118	40.341	1.526
	30	0.458	0.698	1.209	33.576	4.4806
	40	0.442	0.727	1.279	34.284	5.600
	50	0.460	0.738	1.357	29.463	11.990
	60	0.435	0.706	1.422	26.887	15.954
	70	0.453	0.716	1.586	23.701	22.989
	80	0.467	0.702	1.722	23.769	33.415
Lena	20	0.495	0.755	1.066	42.199	1.904
	30	0.512	0.735	1.099	35.522	4.947
	40	0.503	0.756	1.228	31.032	11.917
	50	0.514	0.718	1.359	27.649	23.370
	60	0.549	0.745	1.590	23.577	39.059
	70	0.521	0.717	1.850	22.204	50.050
	80	0.646	0.720	2.064	20.869	58.349

Table 5.
Testing different values of frequency.

Test image	Resolution	Compression time (s)	Decompression time (s)	Compression ratio	MSE	PSNR (dB)
Blonde women	32*32	0.455	0.692	1.223	9.643	32.645
	64*64	2.581	0.698	1.430	13.100	31.678
	128*128	34.634	0.712	1.712	14.165	30.853
	256*256	811.291	0.763	1.741	13.150	32.065
Lena	32*32	0.522	0.704	1.109	5.651	34.695
	64*64	2.193	0.704	1.282	8.722	33.380
	128*128	33.376	0.720	1.486	9.799	32.909
	256*256	732.345	0.763	1.604	9.660	33.115
Cameraman	32*32	0.467	0.832	1.211	3.628	36.328
	64*64	2.608	0.719	1.440	7.743	30.258
	128*128	43.854	0.691	1.658	9.086	31.078
	256*256	732.011	0.977	1.678	9.626	31.095
Living room	32*32	0.558	0.687	1.219	13.475	31.210
	64*64	2.092	0.692	1.355	14.584	31.414
	128*128	26.347	0.710	1.487	14.584	31.322
	256*256	478.219	0.767	1.555	14.138	31.599
Mandrill	32*32	0.598	0.554	1.211	11.936	31.674
	64*64	2.376	0.889	1.422	18.377	30.555
	128*128	22.190	0.732	1.419	18.435	30.582
	256*256	334.636	0.795	1.368	16.404	30.737

Table 6.
Testing the image resolutions.

Figure 3 shows the image of blonde women before and after proposed compression.

In **Figure 4**, we show a cameraman image before and after proposed compression,

Figures 5 and **6** describe separately Lena and living room images before and after applying the proposed compression.

Finally, we conclude our sequence of assessments with mandrill image before and after proposed compression, in **Figure** 7.

As we can observe, the images' quality is very suitable.

And to approve this effect, **Table 6** explores additional quality measure.

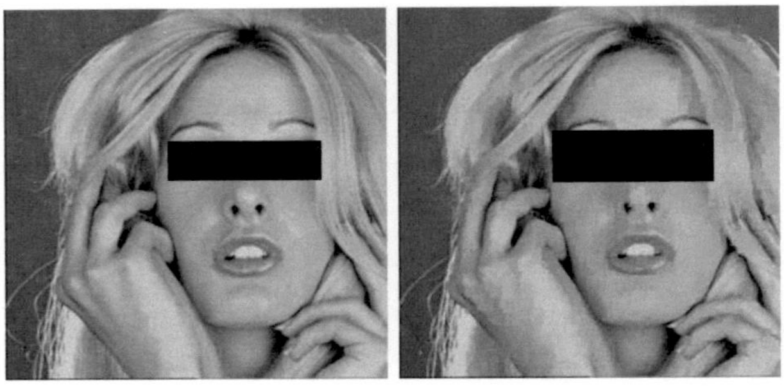

Figure 3.
An image of Compressed and decompressed blond women.

Figure 4.
An image of compressed and decompressed cameramen.

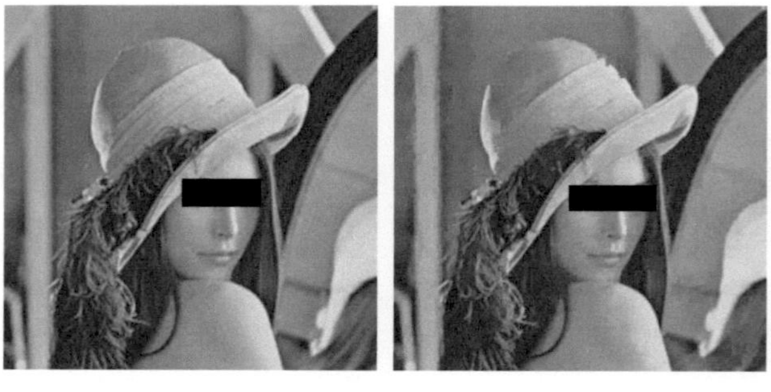

Figure 5.
An image of compressed and decompressed Lena.

The table shows that our approach is significantly sensitive to changing resolutions. It is also clear that the quality of the images is inversely linked to the resolution (as soon as the resolution increases, the quality of the images degrades) which is well proven by the MSE and PSNR measurements.

Figure 6.
An image of compressed and decompressed living room.

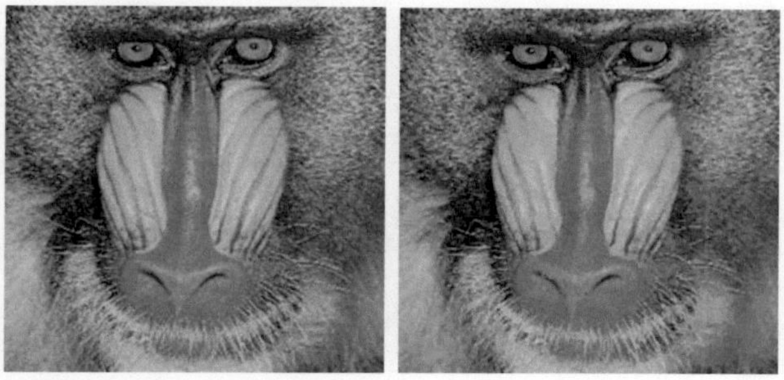

Figure 7.
An image of compressed and decompressed Mandrill.

Test image	Methods	PSNR (dB)	Compression time (s)	Compression ratio
Lena 128*128	BIA	32.909	33.376	1.486
	Suman K. Mitra et al works [7]	30.22	/	1.059
	Vishvas V. Kalunge et al works [42]	/	67	/
Lena 256*256	BIA	33.115	732.345	1.604
	Y. Chakrapani et al.'s works [37]	26.22	2370	1.3
	Exhaustive search	32.69	8400	1.3
	DWSR [44]	25.8212	56.4247	1.56355
	PSO-RCQP [43]	27.089	6.453	1.6392
Cameraman 256*256	BIA	31.095	732.011	1.678
	PSO-RCQP [43]	26.686	268	1.8212
Barbara 128*128	BIA	32.176	21.478	1.312
	Vishvas V. Kalunge et al.'s works [42]	/	66	/

Table 7.
Proposed approach versus other methods.

We note that our proposed method has a remarkable effect on compression ratio and remains relative to the resolution; the bats then show themselves, when the number of blocks is greater (more resolution) by offering a compression ratio greater than that present in the lower resolutions.

On the other hand, the decompression process does not include any complexity compared to the compression process which gives us a decompression time which remains optimized and similar for almost all resolutions.

Finally, and to be in the set of techniques which try to optimize fractal compression, we will draw up a comparison of our approach with certain existing methods. **Table 7** clearly shows the remarkable difference between our method and the others.

At first glance, our proposed method represents new work which has led to satisfactory results. It proportionally retains the quality offered after compression. Reduced time remains very satisfactory, and the compression rate is better than that offered by most of the methods below.

8. Conclusion

In this chapter, we explained the possibility of optimizing fractal image compression using bioinspired metaheuristics. We focused on the application and, for the first time, recent metaheuristics which are, respectively, the wolf pack algorithm and the bat-inspired algorithm in order to improve the performance of fractal image compression. The results demonstrate the efficiency of the algorithms considered.

Compared to other optimization metaheuristics, our approaches offer better results in many aspects, mainly the encoding and decoding time, the size, and the quality.

In addition, proposed approaches demonstrate the ability of these techniques to manage image compression.

Conflict of interest

No potential conflict of interest was reported by the authors.

Thanks

I would to thank the persons that have participated in this work.

Author details

Rafik Menassel
Department of Mathematics and Computer Sciences, Larbi Tebessi University, Tebessa, Algeria

*Address all correspondence to: r.menassel@univ-tebessa.dz

References

[1] Galabov M. Fractal image compression. In: Proceedings of 4th International Conference on Computer Systems and Technologies e-Learning – CompSysTech 2003;**43**(6):2003. pp. 320-326

[2] Selim A, Hadhoud MM, Salem OM. A comparison study between spiral and traditional fractal image compression. In: Proceedings – The 2009 International Conference on Computer Engineering and Systems, ICCES'09. 2009. pp. 39–44

[3] Thanushkodi KG, Bhavani S. Comparison of fractal coding methods for medical image compression. IET Image Processing. 2013;7(7):686-693

[4] Jeng JH, Tseng CC, Hsieh JG. Study on Huber fractal image compression. IEEE Transactions on Image Processing. 2009;**18**(5):995-1003

[5] Li J, Kuo CCJ. Image compression with a hybrid wavelet-fractal coder. IEEE Transactions on Image Processing. 1999;**8**(6):868-874

[6] Han JHJ. Fast fractal image compression using fuzzy classification. In: 2008 Fifth International Conference on Fuzzy Systems and Knowledge Discovery, 3. 2008. pp. 272-276

[7] Mitra SK, Murthy CA, Kundu MK. Technique for fractal image compression using genetic algorithm. IEEE Transactions on Image Processing. 1998;7(4):586-593

[8] Li J, Yuan D, Xie Q, Zhang C. Fractal image compression by ant colony algorithm. In: 2008 the 9th International Conference for Young Computer Scientists. 2008. pp. 1890-1894

[9] Tseng C-C, Hsieh J-G, Jeng J-H. Fractal image compression using visual-based particle swarm optimization. Image and Vision Computing. 2008;**26**(8):1154-1162

[10] Lu G. Fractal image compression. Signal Processing: Image Communication. 1993;5(4):327-343

[11] Hutchinson J. Fractals and self-similarity. Indiana University Mathematics Journal. 1981;**30**(5):713-747

[12] Barnsley MF, Demko S. Iterated function systems and the global construction of fractals. Proceedings of the Royal Society of London A: Mathematical, Physical and Engineering Sciences. 1817;**1985**(399):243-275

[13] Peitgen H-O, Jürgens H, Saupe D. Chaos and fractals. New Frontiers of Science. New York: Springer Science Business Media; 1992:**1**; ISBN: 978-1-4757-4742-3

[14] Geir EÖ, Skjalg L. Fractal-based image coding with fast decoder convergence. Signal Processing. 1994;**40**(1):105-117

[15] Ho Moon Y, Soon Kim H, Shin Kim Y, et al. A novel fast fractal decoding algorithm. Signal Processing: Image Communication. 1999;**14**(4):325-333. DOI: 10.1016/s0923-5965(98)00016-2

[16] Moon YH, Baek KR, Kim YS, et al. Fast fractal decoding algorithm with convergence criteria. Proceedings of the Society of Photo-Optical Instrumentation Engineers. 1997;**36**(7):1992-1999

[17] Jacquin AE. Image coding based on a fractal theory of iterated contractive image transformations. IEEE Transactions on Image Processing. 1992;**1**(1):18-30

[18] Jacquin AE. Fractal image coding: A review. Proceedings of IEEE. 1993; **81**(10):1451-1465

[19] Thomas L, Deravi F. Region-based fractal image compression using heuristic search. IEEE Transactions on Image Processing. 1995;**4**(6):832-838

[20] Cardinal J. Fast fractal compression of greyscale images. IEEE Transactions on Image Processing. 2001;**10**(1): 159-164

[21] He C, Xu X, Yang J. Fast fractal image encoding using one-norm of normalized block. Chaos, Solitons and Fractals. 2006;**27**(5):1178-1186

[22] Tong CS, Pi M. Fast fractal image encoding based on adaptive search. IEEE Transactions on Image Processing. 2001;**10**(9):1269-1277

[23] Hartenstein H, Saupe D. Lossless acceleration of fractal image encoding via the fast Fourier transform. Signal Processing: Image Communication. 2000;**16**(4):383-394

[24] Zhang C, Zhou Y, Zhang Z. Fast fractal image encoding based on special image features. Tsinghua Science and Technology. 2007;**12**(1):58-62

[25] Truong TK, Jeng JH, Reed IS, et al. A fast encoding algorithm for fractal image compression using the DCT inner product. IEEE Transactions on Image Processing. 2000;**9**(4):529-535

[26] Lin YL, Wu MS. An edge property-based neighborhood region search strategy for fractal image compression. Computers & Mathematics with Applications. 2011; **62**(1):310-318. DOI: 10.1016/j.camwa.2011.05.011

[27] Olamaei J, Niknam T, Gharehpetian G. Application of particle swarm optimization for distribution feeder reconfiguration considering distributed generators. Applied Mathematics and Computation. 2008; **201**(1–2):575-586. DOI: 10.1016/j.amc.2007.12.053

[28] Gandomi AH, Alavi AH. Multi-stage genetic programming: A new strategy to nonlinear system modeling. Information Sciences (Ny). 2011;**181**(23):5227-5239. DOI: 10.1016/j.ins.2011.07.026

[29] Blum C, Merkle D. Swarm intelligence: Introduction and applications. In: Natural Computing Series. Berlin, Heidelberg: Springer-Verlag; 2008. ISBN: 978-3-540-74088-9. DOI: 10.1007/978-3-540-74089-6_2

[30] Felix T, Chan S, Tiwari MK. Swarm Intelligence, Focus on Ant and Particle Swarm Optimization, (Ed.), ISBN: 978-3-902613-09-7. Rijeka: IntechOpen; 2007

[31] Wu HS, Zhang FM. Wolf pack algorithm for unconstrained global optimization. In: Mathematical Problems in Engineering. Hindawi Publishing Corporation. Vol. 2014. 2014. Available from: http://dx.doi.org/10.1155/2014/465082. Article ID: 465082

[32] Yang XS. A new metaheuristic bat-inspired algorithm. In: González JR, Pelta DA, Cruz C, Terrazas G, Krasnogor N, editors. Nature Inspired Cooperative Strategies for Optimization (NICSO). Studies in Computational Intelligence. Vol. 284. Berlin, Heidelberg: Springer; 2010. pp. 65-74. DOI: 10.1007/978-3-642-12538-6_6k2

[33] Karaboga D. A comparative study of artificial bee Colony algorithm. Applied Mathematics and Computation. 2009; **214**(1):108-132. DOI: 10.1016/j.amc.2009. 03.090

[34] Martinez C. An ACO algorithm for image compression. CLEI Electronic Journal. 2006;**9**(2):1-17. DOI: 10.19153/cleiej.9.2.1

[35] Chakrapani Y, Soundararajan K. Genetic algorithm applied to fractal image compression. ARPN Journal of Engineering and Applied Sciences. 2009;**4**(1):53-58

[36] Xing-yuan W, Fan-ping L, Shu-guo W. Fractal image compression based on spatial correlation and hybrid genetic algorithm. Journal of Visual Communication and Image Representation. 2009;**20**(8):505-510. DOI: 10.1016/j.jvcir.2009.07.002

[37] Chakrapani Y, Soundararajan K. Implementation of fractal image compression employing particle swarm optimization. The World Journal of Modelling and Simulation. 2010;**6**(1): 40-46

[38] Al-Bundi SS, Al-Saidi NMG, Al-Jawari NJ. Crowding optimization method to improve fractal image compressions based iterated function systems. International Journal of Advanced Computer Science and Applications. 2016;**7**(7):392-401

[39] Al-Saidi NMG, Al-Bundi SS, Al-Jawari NJ. An improved harmony search algorithm for reducing computational time of fractal image coding. Journal of Theoretical and Applied Information Technology. 2017;**95**(8):1669-1678

[40] Menassel R, Nini B, Mekhaznia T. An improved fractal image compression using wolf pack algorithm. Journal of Experimental & Theoretical Artificial Intelligence. 2018;**30**(3):429-439. DOI: 10.1080/0952813X.2017.1409281

[41] Menassel R, Gaba I, Titi K. Introducing BAT inspired algorithm to improve fractal image compression. International Journal of Computers and Applications. 2019. DOI: 10.1080/ 1206212X.2019.1638631

[42] Kalunge VV, Varunakshi B. Time optimization of fractal image compression by using genetic algorithm. International Journal of Engineering Research & Technology. 2012;**1**(10):1-5

[43] Wang X-Y, Zhang D-D, Wei N. Fractal image coding algorithm using particle swarm optimization and hybrid Quadtree partition scheme. IET Image Processing. 2015;**9**(2):153-161

[44] Wang X-Y, Zhang D-D. Discrete wavelet transform-based simple range classification strategies for fractal image coding. Nonlinear Dynamics. 2014; **75**(3):439-448

Chapter 2

Fractal Analysis for Time Series Datasets: A Case Study of Groundwater Quality

Sanjeev Kimothi, Asha Thapliyal and Narendra Singh

Abstract

Fractal dimension (FD) is a highly used mathematical tool to measure long-term memory of time series dataset in various research areas and also applied in chaos theory and fractal and spectral analysis. FD analysis has been applied in various disciplines, e.g., from biophysics, hydrology to computer networking. In developing countries like India, the water quality parameter characterization is very much challenging due to the increase of the contaminated substances in groundwater. In view of health issues and drinking water standards, water quality assurance is a requisite on the region basis. In order to quantify the same, a numerical index known as water quality index (WQI) well adopted by worldwide researchers has been recognized for its significance and applicability for water characterization. Further, the water quality parameters, such as turbidity, chloride, ferrous (Fe), nitrate, pH, calcium (Ca), magnesium (Mg) fluoride (Fl), total dissolved salts (TDS), alkalinity, hardness, and sulfate, could significantly improve the understanding through statistical and fractal modeling. Especially in the high mountainous regions of the Himalayas where there is scarcity of observed dataset, the predictability estimation will be highly applicable in WQI modeling. In the current study, statistical relationship among the sample datasets is obtained by regression equation, coefficient of correlation, Hurst exponent, and FD and probability index between water parameters for Tehri District. It is concluded that the fractal analysis is a better statistical and mathematical tool to calculate water quality indices. Fractal analysis among the various parameters suggested that the water samples are good for drinking and the health.

Keywords: water quality index, fractals, Hurst exponent, Himalayan region

1. Introduction

The Himalayan region has large river basins, and this categorizes it as having a dynamic hydrology which required scientific approaches to study the water resource management and planning at regional basis. Due to large-scale developmental activities along the river basin, the hydrology of the region is affected and hence the water quality. These development activities impact overall water balance of the region, which has a negative effect on various environmental factors like flora, fauna, soil, air, drinking water, and ultimately the human health. Water is a valuable natural resource which contains various suspended substances of organic matter and minerals.

Huge urbanization and drilling activities affect the land transformation, and hence surface water pollution impacts the quality of drinking water [1, 2]. In the indication of the water quality standards and the estimations of constituent concentrations, the laboratory processes and statistical methods are required, and worldwide researchers used the statistical analysis such as multiple linear regressions methods [3–6]. Researches emphasize to examine the physical properties of surface water, and the constituent concentrations can be used for the assessment of water quality situation or water balance with analyzing the parameter deviation and water quality standards [7–11]. Modeling of surface water quality is carried out that assesses numerical indices using earth observation datasets and laboratory methods such as X-ray diffraction technique over Allahabad district, India [12]. It is observed that the statistical methods are capable of comparing the numerical indices. Contaminated quantity separation in the groundwater samples were also studied with numerical index approach and found to be significant for water parameter characterization [13, 14]. Remote sensing and GIS-based approaches along with ground-based dataset have also been applied to study the water quality parameters [15, 16].

Fractal dimension (FD) and laboratory methods together have been used to study the water quality, and the indices are calculated as the weighted average of all observations of interest [17]. Fractal theory has been widely applied on diverse types of datasets in hydrology, geophysics, and climate as well as in other research areas to identify the patterns in time series datasets for describing the irregular and complex behaviors of dynamic systems [18–21]. The rainfall spatiotemporal variations are analyzed for flood seasons in China during 1958–2013 using Hurst exponent and concluded that the rainfall trends will persist in the future also having implications for the ecological restoration and farming operations [22]. Fractal approach has been applied to estimate the climatic indices for climatic variables (pressure, temperature, and rainfall) in the Himalayan foothill region [23]. Fractal dimension demonstrated significant variations from station to station with the values relatively closer to unity at high-altitude sites indicating better climate predictability than that of those over the low-altitude stations in the Himalayan foothills.

In this investigation fractal and statistical analysis is carried out to establish relationships among water parameters such as turbidity, chloride, ferrous, nitrate, pH, calcium, magnesium, fluoride, total dissolved salts (TDS), alkalinity, hardness, and sulfate and to get the significant understanding of WQI. Fractal analysis improves the understanding of WQI especially in the mountainous regions of the Himalayas where 3D models show limitations in resolving the highly complex geographical topography. Despite the advantage with statistical modeling to inherit the effects of terrain and correlated variations among various meteorological parameters, comprehensive investigations of such statistical relationships among observed water quality parameters are lacking over the Himalayan foothills and needs to be studied in detail.

This study intends to carry out the statistical and fractal analysis of groundwater parameters to establish the relationship among the various indices and to understand the behavior of water quality indices (WQI) with the predictability index (PI), Hurst exponent (H), and fractal dimension (D). This study may offer the basic understanding of the WQI of different water parameters regarding the regional hydrogeochemical processes with the laboratory testing methods.

2. Study region and observational dataset

The state of Uttarakhand located in the lap of the central Himalayan region has been identified as a hotspot of anthropogenic stress and one of the most vulnerable regions for climate-mediated risks. The region provides water resources supporting millions of people in South Asia. During the last few decades, the central Himalayan region is observing the cascading effects of the climate change including rise in temperature, receding of glaciers, erratic precipitation patterns, etc. [24]. In the aforementioned scenario, the Himalayan region is receiving global scientific attention for glacier and water resource studies. However, available climate models often have limitation in resolving the highly complex geographical topography in this region which directly or indirectly impacts the water balance over the region. Hence, timely and accurate relationship of water indices using statistical methods inheriting the relationship among water parameters complements the understanding of the available water in the fragile ecosystem of the state of Uttarakhand.

The present study is carried out over the Tehri District of Uttarakhand. The detailed study area map (**Figure 1**) shows the location of the main rivers and drainages along the water quality parameters collected. The observations of water

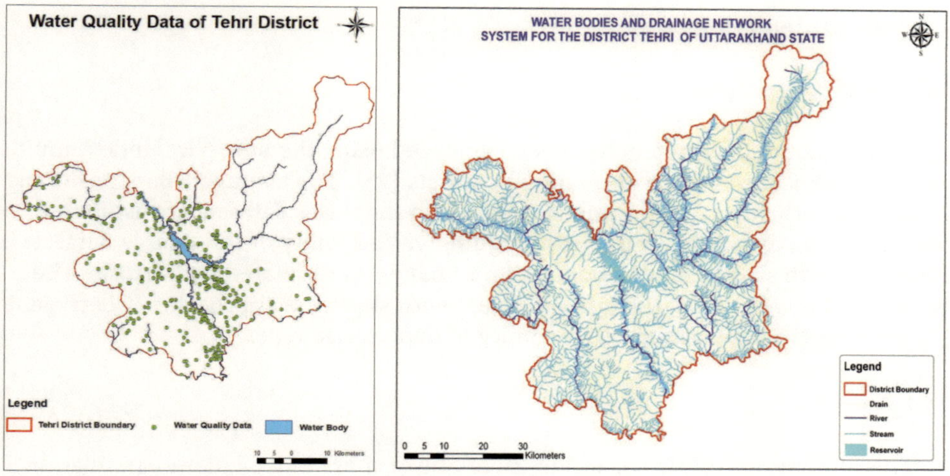

Figure 1.
(a) Data collection points and (b) drainage map of study area.

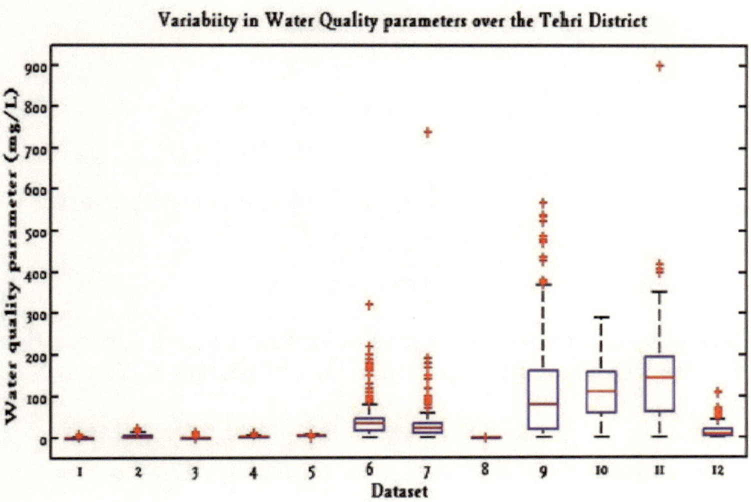

Figure 2.
Box plot showing the water quality parameters over the study site.

quality parameters used for the analysis are obtained through the website of the Ministry of Drinking Water & Sanitation, New Delhi, India.

In this study, we compute the Hurst exponent, fractal dimensions, and the predictability index (PI) of water quality parameters such as (1) turbidity, (2) chloride, (3) ferrous, (4) nitrate, (5) pH, (6) calcium, (7) magnesium, (8) fluoride, (9) total dissolved salts (TDS), (10) alkalinity, (11) hardness, and (12) sulfate, at high-altitude Tehri stations in the Himalayan foothills using the fractal theory. **Figure 2** shows the box plot of all the aforementioned 12 water parameters obtained for the study site. The irregular pattern in the WQI can be used in prediction purposes by analyzing its dynamic flow (i.e., chaotic, random, or deterministic structural pattern). Proper identification, classification, and mapping of water parameters of high-intensive and complex nature require frequent monitoring of these datasets especially in the context of drinking water.

3. Methodology

3.1 Statistical analysis

Water quality parameters have been analyzed using the numerical index, multivariate statistics, and earth observation datasets [25]. The average value, positional average, and the maximum frequency values in the series datasets are estimated with mean, median, and mode correspondingly. Variability of the sample datasets is measured with standard deviation, and peakedness is estimated by kurtosis. The symmetry between data points is estimated with skewness approaches. Coefficient of variation gives the extent of variability of data in a sample.

3.1.1 Regression analysis

This analysis examines the influence of one or more independent variables on a dependent variable. The regression equation with dependent variable Y and independent variable X is represented as:

$$Y = m_{yx}X + C$$

where C is a constant of integration [5]:

$$\text{Regression coefficient} = m_{yx} = r * \left({}^{\sigma_y}\!/_{\sigma_x}\right)$$

$$\text{Correlation coefficient} = r = \frac{[\Psi(XY) - \Psi(X)E(Y)]}{\left[\left(\Psi(X^2) - \Psi(X)^2\right)\left(\Psi(Y^2) - \Psi(Y)^2\right)\right]^{1/2}} = \left[\frac{COV(X, Y)}{\sigma_Y \sigma_X}\right]$$

where σ_y and σ_x are standard deviation of variables Y and X, respectively, and $\Psi(X)$, $\Psi(Y)$, and $\Psi(XY)$ are the expected value of variables X, Y, and XY, respectively.

3.2 Fractal dimension (FD)

The term fractal comes from the Latin word fractus means "fraction or broken"; the basic concept lies in the fact that fractals have a large degree of self-similarity within themselves which was coined by Benot Mandelbrot in 1975. Fractals are characterized by self-similarity property having similar characteristics when analyzed over a large range of scales, and individually a single entity will have similar characteristics to that of the whole fractal [26, 27]. Fractal dimension estimation from a fractal set has various methods due its simplicity and automatic computability. The box counting is one of the major categories of fractal analysis and the most used technique to analyze image features such as texture segmentation, shape classification, and graphic analysis in many fields [28, 29]. The variance and spectral methods are two other major categories of fractal dimension analysis of a time series that recognize the determinism and randomness in data [30]. To study the naturally complex features such as coastlines, river boundaries, mountains, and clouds, the fractal dimension analysis has also provided a mathematical model as a fractal geometry [31, 32]. The glacial and fluvial morphologies are distinguished by using an automated approach (i.e., multifractal). In previous study, a multifractal detrended fluctuation analysis (MFDFA) has been carried out to estimate the variation of elevation profile of glacial and fluvial landscapes [33]. It has been observed that glacial landscapes reveal more complex structure than that of the fluvial landscapes as indicated by fractal parameters, such as degree of multifractality, asymmetry index, etc. The basic definition of fractal dimension is the Hausdorff dimension; however, box counting or box dimension is another popular definition which is easy to calculate.

3.3 The Hurst exponent

Hurst exponent (H) is used as a measure of long-term memory of time series and a real-valued time series defined as the exponent in the asymptotic scaling relation [30, 34]. The Hurst exponent and fractal dimension are also directly related to each other and indicate the roughness of a surface. The Hurst exponent's value lies in a time series as persistent ($0.5 < H \leq 1$) or anti-persistent ($0 \leq H < 0.5$), and when the data are not intercorrelated, then H = 0.5 which implies that the series is unpredictable. This approach is used in various complex engineering fields as it provides statistical self-similarity relationship.

In terms of asymptotic scaling relation, the Hurst exponent of real-valued time series is defined as:

$$\left\langle \frac{R(n)}{S(n)} \right\rangle = Cn^H, \text{as}'n'\text{approaches to infinity} \tag{1}$$

where C is a constant, angular brackets $\langle \cdots \rangle$ denote expected value, S(n) is the standard deviation of the first "n" data of the series$\{X_1, X_2, \cdots, X_n\}$, and R(n) is their range:

$$R(n) = \max\{X_1, X_2, \cdots, X_n\} - \min\{X_1, X_2, \cdots, X_n\}$$

The Hurst exponent H is calculated from rescaled range technique and can also be computed from wavelet method for the time series$\{X_1, X_2, \cdots, X_n\}$.

3.3.1 Estimate of the Hurst exponent: Wavelet approach

If f(t) is a self-affine random process, "t" a position parameter (time or distance), $a > 0$ is a scale (dilatation) parameter, w(t) is a mother wavelet, and

$$w_{t,a}(t') = \frac{1}{\sqrt{a}} w\left(\frac{t'-t}{a}\right)$$

is its shifted, dilatated, and scaled version, then the continuous wavelet transform of f(t) is defined as:

$$W(t,a) = \frac{1}{\sqrt{a}} \int_{-\infty}^{\infty} w_{t,a}(t') f(t') dt' \quad (2)$$

If the time series f(t) is self-affine, the variance of $W(t,a)$ will scale with the dilatation parameter asymptotically as:

$$V(a) = \langle W^2 \rangle - \langle W \rangle^2 \propto a^{\delta} \quad (3)$$

When the exponent "δ" is between −1 and 3 (i.e., $-1 \leq \delta \leq 3$), the Hurst exponent is defined as:

$$H_w = \begin{cases} \frac{\delta+1}{2} & if \quad -1 \leq \delta < 1 \quad \text{(FGN)} \\ \frac{\delta-1}{2} & if \quad 1 \leq \delta \leq 3 \quad \text{(FBM)} \end{cases} \quad (4)$$

where FGN is the fractal Gaussian noise and FBM is the fractional Brownian motion. The Hurst exponent is linked with fractal dimension (D) and defined as:

$$\mathrm{H} = 2 - \mathrm{D} \quad (5)$$

Now the climate predictability index is given as:

$$PI = 2|D - 1.5| = 2|0.5 - H| \quad (6)$$

If PI is close to zero, climate is unpredictable. The closer the PI to 1, the more predictable the climate is.

4. Results and discussion

To distinguish the fresh and contaminated water and establish relationship between the parameters have become a major concern for environmentalists and health workers. And due to increased levels pollutants, it is very challenging for municipal authorities to make availability of clean drinking water especially in developing countries. The statistical relationship of water models depends on the dynamics of climatic as well as soil parameters and thermodynamic processes among the surface water parameters. The established statistical relationship among the various water quality parameters is shown in **Table 1**, which suggested that the variation among these parameters occurs due to variability in the originating environment and is affected by terrain conditions by which it flows down. In dynamic systems, this kind of response generates irregularity, which may show a random pattern of certain type. **Figure 1** shows the box plot of the 12 water quality parameters over the study area. **Table 2** shows the Hurst exponent (H) estimated through standard wavelet techniques and compared with regression equation, and coefficient of each water parameters shows whether they have Brownian time series (or true random walk) behavior with the other related parameters or not. The summary

Sr. No	Parameters	Mean	Median	Mode	SD	Skewness	Kurtosis	Coefficient of variation
1	Turbidity	1.619	1.135	1.000	1.504	1.747	3.134	0.929
2	Chloride	3.114	0.000	3.983	5.009	1.528	1.387	1.609
3	Ferrous	0.266	0.190	0.000	1.031	11.713	150.451	3.881
4	Nitrate	2.663	2.200	0.000	2.097	1.011	1.165	0.787
5	pH	7.189	7.200	0.251	0.691	−7.377	74.926	0.096
6	Calcium	40.656	33.100	0.000	39.952	2.841	11.549	0.983
7	Magnesium	30.725	22.000	0.000	50.368	9.567	127.310	1.639
8	Fluoride	0.283	0.240	0.000	0.266	0.768	−0.157	0.941
9	TDS	107.008	80.150	55.000	116.571	1.781	3.376	1.089
10	Alkalinity	115.239	110.000	1.272	63.728	0.331	−0.683	0.553
11	Hardness	136.661	144.500	95.000	102.589	1.494	8.679	0.751
12	Sulfate	14.564	8.800	0.000	17.161	2.400	7.224	1.178

Table 1.
Statistical analysis of groundwater parameters at Tehri District, Uttarakhand.

Y	Parameters-X	Regression equation	r^2	H	D (Fractal)	PI
Turbidity	Chloride	y = −0.39645*x + 3.7212	0.014385	0.83198	1.168	0.66396
	Ferrous	y = 0.12408*x + 0.065435	0.032746	1.5146	0.48536	2.0293
	Nitrate	y = 0.22208*x + 2.3121	0.02548	0.84779	1.1522	0.69558
	pH	y = 0.010597*x + 7.1721	0.000532	1.2259	0.77415	1.4517
	Calcium	y = 3.6487*x + 34.8722	0.018921	1.2706	0.72942	1.5412
	Magnesium	y = 2.8703*x + 26.1718	0.007351	1.2307	0.76928	1.4614
	Fluoride	y = 0.054962*x + 0.19432	0.097003	0.93987	1.0601	0.87973
	TDS	y = 15.5427*x + 81.9829	0.040221	0.85579	1.1442	0.71158
	Alkalinity	y = 13.4538*x + 93.5935	0.10097	0.82012	1.1799	0.64025
	Hardness	y = 12.7963*x + 116.0534	0.035196	1.2005	0.7995	1.401
	Sulfate	y = 0.8647*x + 13.2096	0.005754	0.90684	1.0932	0.81367
Chloride	Turbidity	y = −0.036284*x + 1.7326	0.014385	0.99887	1.0011	0.99774
	Ferrous	y = 0.0073198*x + 0.2440	0.001245	1.6072	0.39276	2.2145
	Nitrate	y = −0.094225*x + 2.9621	0.050117	0.98909	1.0109	0.97818
	pH	y = 0.017067*x + 7.1368	0.015072	1.1173	0.88266	1.2347
	Calcium	y = −0.18703*x + 41.362	0.000543	1.3236	0.6764	1.6472
	Magnesium	y = −1.4045*x + 35.148	0.019231	1.3891	0.61086	1.7783
	Fluoride	y = 0.0032535*x + 0.2733	0.003714	1.0405	0.95945	1.0811
	TDS	y = 5.7221*x + 89.5595	0.059563	0.9857	1.0143	0.9714
	Alkalinity	y = 3.5154*x + 104.5777	0.075324	0.92132	1.0787	0.84263
	Hardness	y = 3.3113*x + 126.6005	0.02575	1.3457	0.65428	1.6914
	Sulfate	y = 1.5794*x + 9.7489	0.20972	1.0341	0.96585	1.0683

Y	Parameters-X	Regression equation	r^2	H	D (Fractal)	PI
Ferrous	Turbidity	y = 0.2639*x + 1.5505	0.032746	0.40358	1.5964	0.19285
	Chloride	y = 0.1701*x + 3.0332	0.001245	0.3567	1.6433	0.2866
	Nitrate	y = 0.21107*x + 2.6158	0.010822	0.52792	1.4721	0.055837
	pH	y = 0.018421*x + 7.1844	0.000756	0.80777	1.1922	0.61553
	Calcium	y = 4.7305*x + 39.5253	0.014954	0.66518	1.3348	0.33036
	Magnesium	y = 1.1981*x + 30.5048	0.000602	0.63682	1.3632	0.27364
	Fluoride	y = 0.012348*x + 0.28012	0.002302	0.50357	1.4964	0.007137
	TDS	y = −1.9056*x + 107.6835	0.000284	0.34176	1.6582	0.31647
	Alkalinity	y = 4.7751*x + 114.1275	0.005981	0.32296	1.677	0.35407
	Hardness	y = 5.5291*x + 135.3207	0.00309	0.50895	1.491	0.017901
	Sulfate	y = 2.2747*x + 14.0048	0.018721	0.42956	1.5704	0.14088
Nitrate	Turbidity	y = 0.11474*x + 1.3143	0.02548	0.68125	1.3188	0.3625
	Chloride	y = −0.53189*x + 4.4998	0.050117	0.662	1.338	0.324
	Ferrous	y = 0.051273*x + 0.12956	0.010822	1.5921	0.40791	2.1842
	pH	y = 0.049051*x + 7.0583	0.022056	1.3583	0.64173	1.7165
	Calcium	y = 5.7087*x + 25.5324	0.089649	1.1608	0.83923	1.3215
	Magnesium	y = 7.2362*x + 11.4886	0.090436	1.2002	0.7998	1.4004
	Fluoride	y = 0.033377*x + 0.19422	0.069242	0.92495	1.0751	0.84989
	TDS	y = −12.2844*x + 140.0001	0.048631	0.62589	1.3741	0.25179
	Alkalinity	y = 0.37421*x + 114.4004	0.000151	0.58893	1.4111	0.17786
	Hardness	y = 3.2928*x + 127.996	0.004511	0.91075	1.0892	0.82151
	Sulfate	y = −0.41289*x + 15.7144	0.002539	0.80191	1.1981	0.60382
pH	Turbidity	y = 0.050188*x + 1.2601	0.000532	0.37731	1.6227	0.24538
	Chloride	y = 0.88314*x + −3.2706	0.015072	0.28645	1.7135	0.4271
	Ferrous	y = 0.041021*x + −0.02835	0.000756	0.9331	1.0669	0.8662
	Nitrate	y = 0.44965*x + −0.56059	0.022056	0.52027	1.4797	0.040543
	Calcium	y = 4.7201*x + 6.8518	0.006686	0.66003	1.34	0.32006
	Magnesium	y = 4.6756*x + −2.7902	0.004119	0.65269	1.3473	0.30538
	Fluoride	y = 0.029938*x + 0.068174	0.006077	0.5019	1.5008	0.001615
	TDS	y = 1.682*x + 95.0834	9.95E-05	0.2787	1.7213	0.4426
	Alkalinity	y = 3.1687*x + 92.6195	0.001183	0.2732	1.7268	0.4536
	Hardness	y = 2.6385*x + 117.8255	0.000316	0.46067	1.5393	0.078657
	Sulfate	y = 2.1465*x + −0.82098	0.007486	0.39596	1.604	0.20807
Calcium	Turbidity	y = 0.0051856*x + 1.4094	0.018921	0.65827	1.3417	0.31654
	Chloride	y = −0.0029044*x + 3.197	0.000543	0.57117	1.4288	0.14234
	Ferrous	y = 0.0031612*x + 0.13763	0.014954	1.2934	0.70663	1.5867
	Nitrate	y = 0.015704*x + 2.0315	0.089649	0.7484	1.2516	0.4968
	pH	y = 0.0014165*x + 7.1316	0.006686	1.111	0.88902	1.222
	Magnesium	y = 0.69708*x + 2.3928	0.30508	0.97651	1.0235	0.95303
	Fluoride	y = 0.00093543*x + 0.24526	0.019771	0.74791	1.2521	0.49582

Y	Parameters-X	Regression equation	r^2	H	D (Fractal)	PI
	TDS	y = −0.038294*x + 108.737	0.000172	0.56384	1.4362	0.12768
	Alkalinity	y = 0.62046*x + 90.0942	0.1511	0.53726	1.4627	0.074521
	Hardness	y = 0.67582*x + 109.2303	0.069075	0.82607	1.1739	0.65214
	Sulfate	y = 0.089831*x + 10.9473	0.043691	0.67248	1.3275	0.34495
Magnesium	Turbidity	y = 0.0025611*x + 1.5419	0.007351	0.57726	1.4227	0.15451
	Chloride	y = −0.013693*x + 3.5007	0.019231	0.5427	1.4573	0.0854
	Ferrous	y = 0.0005026*x + 0.2510	0.000602	1.121	0.87899	1.242
	Nitrate	y = 0.012498*x + 2.2868	0.090436	0.70056	1.2994	0.40112
	pH	y = 0.0008809*x + 7.1622	0.004119	1.01946	1.0054	0.98923
	Calcium	y = 0.43765*x + 27.2961	0.30508	0.88407	1.1159	0.76813
	Fluoride	y = 0.0011794*x + 0.2470	0.050062	0.69593	1.3041	0.39186
	TDS	y = −0.16172*x + 112.1606	0.00488	0.52529	1.4747	0.050579
	Alkalinity	y = 0.16349*x + 110.361	0.01671	0.5019	1.5041	0.008196
	Hardness	y = 0.94027*x + 107.8115	0.21297	0.8051	1.1949	0.61019
	Sulfate	y = 0.009879*x + 14.3066	0.000842	0.62993	1.3701	0.25985
Fluoride	Turbidity	y = 1.7649*x + 1.1207	0.097003	0.82823	1.1718	0.65646
	Chloride	y = 1.1415*x + 2.7551	0.003714	0.76375	1.2362	0.5275
	Ferrous	y = 0.18644*x + 0.21372	0.002302	1.6654	0.33457	2.3309
	Nitrate	y = 2.0745*x + 2.0841	0.069242	1.0143	0.98566	1.0287
	pH	y = 0.20299*x + 7.1318	0.006077	1.4292	0.5708	1.8584
	Calcium	y = 21.1354*x + 34.7963	0.019771	1.2721	0.72787	1.5443
	Magnesium	y = 42.446*x + 18.7946	0.050062	1.3075	0.6925	1.615
	TDS	y = −77.0811*x + 129.021	0.030805	0.74411	1.2559	0.48822
	Alkalinity	y = 63.845*x + 97.3061	0.070809	0.7059	1.2941	0.41179
	Hardness	y = 43.9117*x + 124.3496	0.012907	1.0748	0.92515	1.1497
	Sulfate	y = 20.7924*x + 8.7184	0.1036	0.89597	1.104	0.79193
TDS	Turbidity	y = 0.0025877*x + 1.3435	0.040221	1.0328	0.96716	1.0657
	Chloride	y = 0.010409*x + 1.963	0.059563	1.0908	1.0091	0.98171
	Ferrous	y = −0.0001491*x + 0.28255	0.000284	1.548	0.452	2.096
	Nitrate	y = −0.0039588*x + 3.0963	0.048631	0.94004	1.06	0.88008
	pH	y = 5.9129e-05*x + 7.183	9.95E-05	1.0928	0.9072	1.1856
	Calcium	y = −0.004486*x + 41.267	0.000172	1.3135	0.68655	1.6269
	Magnesium	y = −0.030176*x + 34.0583	0.00488	1.3516	0.64839	1.7032
	Fluoride	y = −0.0003996*x + 0.32624	0.030805	1.0191	0.98091	1.0382
	Alkalinity	y = 0.20175*x + 93.7774	0.13638	0.93687	1.0631	0.87375
	Hardness	y = 0.26534*x + 108.3566	0.090894	1.3517	0.64832	1.7034
	Sulfate	y = 0.037752*x + 10.565	0.06587	1.0279	0.97206	1.0559
Alkalinity	Turbidity	y = 0.0075049*x + 0.75479	0.10097	1.1155	0.88449	1.231
	Chloride	y = 0.021427*x + 0.60594	0.075324	1.0438	0.95622	1.0876
	Ferrous	y = 0.0012524*x + 0.12203	0.005981	1.6487	0.35134	2.2973

Y	Parameters-X	Regression equation	r^2	H	D (Fractal)	PI
	Nitrate	y = 0.00040404*x + 2.6254	0.000151	0.98687	1.0031	
	pH	y = 0.0003732*x + 7.1463	0.001183	1.2073	0.79271	1.4146
	Calcium	y = 0.24353*x + 12.683	0.1511	1.4105	0.58949	1.821
	Magnesium	y = 0.10221*x + 19.0295	0.01671	1.4381	0.56193	1.8761
	Fluoride	y = 0.0011091*x + 0.1554	0.070809	1.0896	0.91044	1.1791
	TDS	y = 0.67597*x + 29.1685	0.13638	1.0559	0.94412	1.1118
	Hardness	y = 0.53406*x + 75.1633	0.1099	1.4326	0.56739	1.8652
	Sulfate	y = 0.12922*x + −0.30058	0.23032	1.0907	0.9093	1.1814
Hardness	Turbidity	y = 0.0027505*x + 1.2446	0.035196	0.73752	1.2625	0.47503
	Chloride	y = 0.0077766*x + 2.0148	0.02575	0.6886	1.3114	0.37721
	Ferrous	y = 0.0005587*x + 0.19012	0.00309	1.1735	0.82654	1.3469
	Nitrate	y = 0.0013699*x + 2.4847	0.004511	0.69629	1.3037	0.39259
	pH	y = 0.0001197*x + 7.1729	0.000316	0.91947	1.0805	0.83895
	Calcium	y = 0.10221*x + 26.8046	0.069075	0.97954	1.0205	0.95908
	Magnesium	y = 0.2265*x + −0.15996	0.21297	1.0545	0.9455	1.109
	Fluoride	y = 0.0002939*x + 0.2432	0.012907	0.74932	1.2507	0.49865
	TDS	y = 0.34256*x + 60.3158	0.090894	0.68805	1.312	0.3761
	Alkalinity	y = 0.20579*x + 87.25	0.1099	0.64706	1.3529	0.29411
	Sulfate	y = 0.044992*x + 8.4565	0.072467	0.73569	1.2643	0.47139
Sulfate	Turbidity	y = 0.0066538*x + 1.5236	0.005754	0.93736	1.0626	0.87472
	Chloride	y = 0.13279*x + 1.1384	0.20972	0.89036	1.1096	0.78071
	Ferrous	y = 0.00823*x + 0.14631	0.018721	1.6664	0.33359	2.3328
	Nitrate	y = −0.0061497*x + 2.7619	0.002539	1.0315	0.96846	1.0631
	pH	y = 0.0034876*x + 7.1384	0.007486	1.3298	0.67025	1.6595
	Calcium	y = 0.48637*x + 33.6799	0.043691	1.3417	0.65832	1.6834
	Magnesium	y = 0.085194*x + 29.5794	0.000842	1.3882	0.61179	1.7764
	Fluoride	y = 0.004982*x + 0.21061	0.1036	1.051	0.94905	1.1019
	TDS	y = 1.7448*x + 81.682	0.06587	0.8804	1.1196	0.76081
	Alkalinity	y = 1.7825*x + 89.3566	0.23032	0.82887	1.1711	0.65774
	Hardness	y = 1.6107*x + 113.2606	0.072467	1.2378	0.76216	1.4757

Table 2.
Regression equation, coefficient of correlation, Hurst exponent, fractal dimension, and probability index between water parameters at Tehri District.

of the statistical and fractal analysis is shown in **Table 2**, and each WQI analysis is discussed subsequently.

4.1 Turbidity

The turbidity sample datasets exhibit normal behavior as the mean, median, and mode values are approximately equal. Standard deviation is found to be 1.5 and suggests that the sample data points are close together. The positive skewness (1.747) of data points reveals that the curve is not symmetrical, and the kurtosis value 3.13 shows that the sample datasets are platykurtic. Turbidity has persistent behavior with chloride, nitrate, fluoride, TDS, alkalinity, and sulfate and anti-persistent behavior with ferrous, PH, Ca, Mg, and hardness parameters.

4.2 Chloride

Mean and mode values are in the order of ± 0.5, and thus the data show normal behavior even though the median is 0. High standard deviation (5.009) is observed between sample points. Skewness value (1.52) suggests that the curve is not symmetrical, and the kurtosis value (1.3) is less than 3. Chloride has the Brownian time series (true random walk) behavior with Fl, sulfate, and turbidity parameters. Thus, the curve is platykurtic. Chloride has persistent behavior with turbidity, nitrate, TDS, and alkalinity and anti-persistent behavior with Fe, PH, Ca, Mg, and hardness parameters.

4.3 Ferrous (Fe)

Average, median, and mode values are approximately equal, and thus the data show normal behavior. Standard deviation value (1.031) exhibits that the sample points are close to each other. Skewness value (11.713) suggests that the curve is not symmetrical, and kurtosis value is very large; thus, the curve is not platykurtic. The sample dataset containing heavier outliers and Fe has Brownian time series (True random walk) behavior with nitrate, fluoride, and hardness parameters. It has persistent behavior with pH, Ca, and Mg and anti-persistent behavior with chloride, TDS, alkalinity, and sulfate parameters.

4.4 Nitrate

Mean and median values and standard deviation are approximately equal; thus data exhibit normal behavior. This suggests that the sample data are close to each other. The skewness value (1.011) and kurtosis are less than 3; hence the curve is not symmetrical and platykurtic. Nitrate has persistent behavior with turbidity, chloride, fluoride, TDS, alkalinity, hardness, and sulfate and anti-persistent behavior with Ca, Mg, Fe, and PH parameters.

4.4.1 pH

Average and median are almost same, i.e., 7.189 and 7.20, respectively, whereas the mode of pH is 0.25. These values are approximately equal and hence exhibit the normal behavior. Standard deviation (SD) is 0.691, and skewness is close to 0, and all values are also close to each other; thus pH is symmetrical. The curve is not platykurtic, as kurtosis is very large 74.92. It shows the Brownian time series behavior with fluoride (Fl) parameter; persistent behavior with Ca, Mg, Fe, and nitrate; and anti-persistent performance with different parameters, i.e., turbidity, chloride, TDS, alkalinity, hardness, and sulfate.

4.4.2 Calcium

Mean, median, and mode values are not close to each other; thus the curve does not show normal behavior. High standard deviation ($\sim$40) indicates that the Ca values are very much distributed from each other. It is positively skewed, and the curve is not platykurtic. With few parameters, i.e., turbidity, chloride, TDS, alkalinity, hardness, Mg, Fl, TDS, and sulfate, it shows persistent behavior and anti-persistent behavior with Fe and pH parameters.

4.4.3 Magnesium

Mean and mode values are 30.75 and 22.0, respectively, and median is 0, so the sample dataset are not same, and thus the curve does not show normal behavior. Standard deviation value is high (50.368); thus, the values of Mg are very much distributed with each other. It is positively skewed, and the curve is not platykurtic. Mg has Brownian time series (true random walk) behavior with pH and alkalinity parameters. Mg has persistent behavior with turbidity, chloride, nitrate, Ca, TDS, hardness, Fl, and sulfate and anti-persistent behavior with Fe parameters.

4.4.4 Fluoride

Mean and median values are approximately equal, and thus the curve shows normal behavior. Standard deviation value (0) suggests that the sample data are close to each other, and the skewness and kurtosis value suggest that curve is platykurtic. Fl has Brownian time series (true random walk) behavior with nitrate and hardness parameters. Fl has persistent behavior with turbidity, chloride, TDS, alkalinity, and sulfate and anti-persistent behavior with Fe, pH, Ca, and Mg parameters.

4.4.5 Total dissolved salts (TDS)

Mean, median, and mode values are different; thus, the curve does not follow normal behavior. Standard deviation value is high (116.57); thus the values of TDS are not close to each other. TDS has Brownian time series (true random walk) behavior with turbidity, chloride, pH, Fl, and sulfate parameters. It is negatively skewed and the curve is platykurtic. TDS has persistent behavior with alkalinity, nitrate, and chloride and anti-persistent behavior with Fe, Ca, Mg, and hardness parameters.

4.4.6 Alkalinity

Average, median, and mode (1.27) values are nearly equal to each other, and sample data exhibit normal behavior. High standard deviation (63.72) is observed between the datasets, with the skewness value (0.331) which is close to 0. The kurtosis value is less than 3; thus the curve is symmetrical and platykurtic. It has Brownian time series behavior with chloride, Fl, TDS, and sulfate parameters. Alkalinity has persistent behavior with nitrate and anti-persistent behavior with turbidity, Fe, pH, Ca, Mg, and hardness parameters.

4.4.7 Hardness

The data series does not exhibit normal behavior as the mean and median values are a having large difference with the mode value (95.0). Standard deviation value (102.5) suggests that data are spread out, and skewness values observed to be 1.49; hence the curve is platykurtic. Only with Mg parameter, it has Brownian time series flow. Hardness has persistent behavior with turbidity, chloride, nitrate, PH, Ca, Fl, TDS, alkalinity, and sulfate and anti-persistent behavior with Fe parameters.

4.4.8 Sulfate

Mean and median values are different and mode value is 0. Standard deviation (17.16) reveals that the dataset has distributed form. The skewness value is equal to 2.40 with larger kurtosis value, i.e., 7.22, which indicates that the curve is not symmetrical. It has true random walk flow with Fl and nitrate parameters. Sulfate has persistent behavior with turbidity, chloride, TDS, and alkalinity and anti-persistent behavior with hardness, Fe, PH, Ca, and Mg parameters.

5. Conclusion

The water parameters from different sources in the Tehri District of Uttarakhand have shown the non-platykurtic curve. The analysis of most of the water parameter combinations has shown the Brownian time series behavior with each other. The irregular pattern in the WQI can be used for prediction purposes by deciding if its dynamic follows a chaotic, random, or deterministic structural pattern. Mostly all groundwater variables like turbidity, chloride, iron, nitrate, pH, calcium, magnesium, fluoride, TDS, alkalinity, hardness, sulfate, etc. are affected by each other. The pH of the sample datasets shows the Brownian time series behavior with fluoride (Fl) parameter; persistent behavior with Ca, Mg, Fe, and nitrate; and anti-persistent performance with turbidity, chloride, TDS, alkalinity, hardness, and sulfate. Turbidity, chloride, nitrate, fluoride, TDS, alkalinity, sulfate, and chloride have shown persistent behavior with each other. Fe has persistent behavior with pH, Ca, and Mg, and nitrate has persistent behavior with turbidity, chloride, fluoride, TDS, alkalinity, hardness, and sulfate. pH has persistent behavior with Ca, Mg, Fe, and nitrate. Turbidity, chloride, TDS, alkalinity, hardness, Mg, Fl, TDS, sulfate, and Ca show persistent behavior. Mg has persistent behavior with turbidity, chloride, nitrate, Ca, TDS, hardness, Fl, and sulfate. Fl has persistent behavior with turbidity, chloride, TDS, alkalinity, and sulfate. TDS has persistent behavior with alkalinity, nitrate, and chloride. Alkalinity has persistent behavior with nitrate only. Hardness has persistent behavior with turbidity, chloride, nitrate, pH, Ca, Fl, TDS, alkalinity, and sulfate. Sulfate has persistent behavior with turbidity, chloride, TDS, and alkalinity.

Turbidity and chloride have anti-persistent behavior with Fe, pH, Ca, Mg, and hardness parameters. Fe has anti-persistent behavior with chloride, TDS, alkalinity, and sulfate parameters. Nitrate has anti-persistent behavior with Ca, Mg, Fe, and pH parameters. pH has anti-persistent performance with different parameters, i.e., turbidity, chloride, TDS, alkalinity, hardness, and sulfate. Mg has anti-persistent behavior with Fe parameters only and Ca with Fe and pH parameters. Fl has anti-persistent behavior with Fe, pH, Ca, and Mg parameters. TDS has anti-persistent behavior with Fe, Ca, Mg, and hardness parameters. Alkalinity has anti-persistent behavior with turbidity, Fe, pH, Ca, Mg, and hardness parameters. Hardness has anti-persistent behavior with Fe parameter only, and sulfate has anti-persistent behavior with hardness, Fe, PH, Ca, and Mg parameters.

The persistent behavior is observed among the various indices which reveal that the variations of the water quality parameters are under an acceptable range with each other. This study is focused on the utility of the Hurst exponent, fractal dimension as an analysis tool, and predictability indices (PI) along with regression and coefficient of correlation among the water quality time series data points. It is concluded that the fractal analysis is a better statistical and mathematical tool to calculate water quality indices. Fractal analysis among the various parameters suggested that the water samples are good for drinking and the health.

Acknowledgements

The authors would like to thank the Ministry of Drinking Water & Sanitation, New Delhi, India, for providing the groundwater quality datasets (through website during 2016-17) cited in this study.

The contents and views expressed in this article are the views of the authors and do not necessarily reflect the views of the organizations they belong to.

Author details

Sanjeev Kimothi[1]*, Asha Thapliyal[2] and Narendra Singh[3]

1 Department of Physics, Himalayan School of Science and Technology, Swami Rama Himalayan University (HSST, SRHU), Doiwala Dehradun Uttarakhand, India

2 Water Resource Department, Uttarakhand Space Application Centre (USAC), Dehradun, Uttarakhand, India

3 Aryabhatta Research Institute of Observational Sciences (ARIES), Nainital, Uttarakhand, India

*Address all correspondence to: sanjeevkimothi@gmail.com

References

[1] Amin A, Fazal S, Mujtaba A, Singh SK. Effects of land transformation on water quality of Dal Lake, Srinagar, India. Journal of the Indian Society of Remote Sensing. 2014;**42**(1):119-128. DOI: 10.1007/s12524-013-0297-9

[2] Thakur JK, Diwakar J, Singh SK. Hydrogeochemical evaluation of groundwater of Bhaktapur municipality, Nepal. Environment and Earth Science. 2015;**74**(6):4973-4988. DOI: 10.1007/s12665-015-4514-4

[3] World Health Organization (WHO). Guidelines for Drinking-Water Quality; Recommendations. 3rd ed. Vol. 1. Geneva: World Health Organization; 2006. p. 595. ISBN: 92 4 154696 4

[4] Chenini I, Ben Mammou A, Turki MM. Groundwater resources of a multi-layered aquiferous system in arid area: Data analysis and water budgeting. International journal of Environmental Science and Technology. 2008;**5**(3):361-374

[5] Chenini I, Khemiri S. Evaluation of ground water quality using multiple linear regression and structural equation modeling. International journal of Environmental Science and Technology. 2009;**6**(3):509-519

[6] Mousavi M, Kiani S, Lotfi S, Naeemi N, Honarmand M. Transient and spatial modelling and simulation of polybrominated diphenyl ethers reaction and transport in air, water and soil. International journal of Environmental Science and Technology. 2008;**5**(3):323-330

[7] Vassilis Z, Antonopoulos M, Mitsiou AK. Statistical and trend analysis of water quality and quantity data for the Strymon River in Greece. Hydrology and Earth System Sciences. 2001;**5**(4):679-691

[8] MAM J, Raihan F, Alam JB, Hasanuzzaman S. Regression analysis of ground water quality data of Sunamganj district, Bangladesh. International Journal of Environmental Research. 2008;**2**(3):291-296

[9] Carlson E, Ecker MD. A statistical examination of water quality in two Iowa Lakes. American Journal of Undergraduate Research. 2002;**1**(2):31-45

[10] Korashey R. Using regression analysis to estimate water quality constituents in Bahr El Baqar drain. Journal of Applied Sciences Research. 2009;**5**(8):1067-1076

[11] Psargaonkar A, Gupta A, Devotta S. Multivariate analysis of ground water resources in Ganga–Yamuna Basin (India). Journal of Environmental Science & Engineering. 2008;**50**(3):215-222

[12] Singh SK, Srivastava PK, Singh D, Han D, Gautam SK, Pandey AC. Modeling groundwater quality over a humid subtropical region using numerical indices, earth observation datasets, and X-ray diffraction technique: A case study of Allahabad district, India. Environmental Geochemistry and Health. 2015;**37**(1):157-180. DOI: 10. 1007/s10653-014-9638-z

[13] Rawat KS, Mishra AK, Singh SK. Mapping of groundwater quality using normalized difference dispersal index of Dwarka subcity at Delhi National Capital of India. ISH Journal of Hydraulic Engineering. 2017;**23**(3):229-240. DOI: 10.1080/09715010.2016.1277795

[14] Rawat KS, Tripathi VK, Singh SK. Groundwater quality evaluation using numerical indices: A case study (Delhi, India). Sustainable Water Resources

Management. 2018;**4**:875-885. DOI: 10.1007/s40899-017-0181-9

[15] Singh SK, Srivastava PK, Pandey AC. Fluoride contamination mapping of groundwater in northern India integrated with geochemical indicators and GIS. Water Science and Technology: Water Supply. 2013;**13**(6): 1513-1523. DOI: 10.2166/ws.2013.160

[16] Singh SK, Srivastava PK, Pandey AC. Integrated assessment of groundwater influenced by a confluence river system: Concurrence with remote sensing and geochemical modeling. Water Resources Management. 2013;**27**: 4291-4313. DOI: 10.1007/s11269-013-0408-y

[17] Parmar KS, Bhardwaj R. Water quality index and fractal dimension analysis of water parameters. International journal of Environmental Science and Technology. 2013;**10**: 151-164. DOI: 10.1007/s13762-012-0086-y

[18] Rehman S. Study of Saudi Arabian climatic conditions using Hurst exponent and climatic predictability index. Chaos, Solitons & Fractals. 2009; **39**(2):499-509

[19] Mandelbrot BB, Wallis JR. Noah, Joseph and operational hydrology. Water Resources Research. 1968;**4**: 909-918

[20] Mandelbrot BB, Wallis JR. Some long run properties of geophysical records. Water Resources Research. 1969;**5**:228

[21] Men B, Liang C, Zhao X. Chaotic analysis on precipitation times series of Sicuani middle part in upper region of Yangtze. Nature and Science. 2004;**2**: 74-78

[22] He Y, Mu X, Gao P, Zhao G, Wang F, Sun W, et al. Spatial variability and periodicity of precipitation in the middle reaches of the Yellow River, China. Advances in Meteorology. 2016; **2016**. Article ID 9451614. DOI: 10.1155/2016/9451614

[23] Kimothi S, Kumar A, Thapliyal A, Ojha N, Soni VK, Singh N. Climate predictability in the Himalayan foothills using fractals. Mausam. 2019;**70**(2): 357-362

[24] Singh N, Solanki R, Ojha N, Janssen RHH, Pozzer A, Dhaka SK. Boundary layer evolution over the Central Himalayas from radio wind profiler and model simulations. Atmospheric Chemistry and Physics. 2016;**16**:10559-10572. DOI: 10.5194/acp-16-10559-2016

[25] Singh SK, Singh P, Gautam SK. Appraisal of urban lake water quality through numerical index, multivariate statistics and earth observation data sets. International Journal of Environmental Science and Technology. 2016;**3**(2):445-456. DOI: 10.1007/s13762-015-0850-x

[26] Mandelbrot BB, Van Ness JW. Fractional Brownian motion, fractional noises and applications. SIAM Review. 1968;**10**:422-427

[27] Hurst HE, Black RP, Simaika YM. Long-Term Storage: An Experimental Study. London: Constable Co. Ltd.; 1965

[28] Rangarajan G, Sant DA. A climatic practicability index and its applications. Geophysical Research Letters. 1997;**24**: 1239-1242

[29] Turcotte DL. Fractals and Chaos in Geology and Geophysics-Second Edition. Cambridge: Cambridge University Press; 1997

[30] Rehman S, Siddiqi AH. Wavelet based Hurst exponent and fractal dimensional analysis of Saudi climatic dynamics. Chaos, Solitons & Fractals. 2009;**40**(3):1081-1090

[31] Mandelbrot BB. The Fractal Geometry of Nature. New York, NY, USA: W. H. Freeman and Company; 1982. p. 656

[32] Rodriguez-Iturbe I, Marani M, Rigon R, Rinaldo A. Self organized river basin landscape: Fractal and multifractal characteristics. Water Resources Research. 1994;**30**:3531-3539

[33] Dutta S. Decoding the morphological differences between Himalayan glacial and fluvial landscapes using multifractal analysis. Scientific Reports. 2017;7:11032. DOI: 10.1038/s41598-017-11669-0

[34] Peitgen HO, Jiigens H, Sample D. Chaos and Fractals New Frontiers of Science. Heidelberg: Springer; 1992

Chapter 3

Establishing the Downscaling and Spatiotemporal Scale Conversion Models of NDVI Based on Fractal Methodology

Haijun Luan

Abstract

Scale effect is a crucial scientific problem in quantitative remote sensing (RS), and scholars attempt to solve it with scale conversion models, which can characterize the numerical relationship of RS land surface parameters at different resolutions (scales). As a significant land surface parameter, scale conversion of normalized difference vegetation index (NDVI) has been studied for a long time. Therefore, taking NDVI as an example, the development of scaling research is described and analyzed in the paper, and based on fractal theory, the development trends are discussed for land surface parameters in quantitative remote sensing. These are our conclusions: it will be the new trend to establish downscaling models based on fractal theory for land surface parameters in quantitative remote sensing; additionally, it still is the hotspot to establish spatiotemporal scale conversion models for land surface parameters in quantitative remote sensing in the future, and addressed on that, the multi-fractal scaling methodology is proposed, and its availability is analyzed in the paper, which presents significant potential.

Keywords: remote sensing (RS), normalized difference vegetation index (NDVI), scaling, fractal, iterated function system (IFS), multi-fractal

1. Introduction

The scale problem is one of the important and fundamental problems of quantitative remote sensing [1–3]. Scholars have studied the scale effects of different remote sensing (RS) land surface parameters. The study of scale effect is conducive to the synergistic use of RS data of different spatial and temporal resolutions (scales) to solve the problem that "massive" RS images cannot be fully utilized and has important application potential and scientific research value [3]. In view of the spatiotemporal characteristics of the ground objects, the RS land surface parameters not only have spatial scale effects but also have temporal scale effects. Scholars have conducted extensive and in-depth research on the scale effect of land surface parameters, which includes the mechanism, manifestation, effect analysis, and solution of scale effects. The author has previously discussed it in detail [4]. Based on the above research aspects, scale conversion as a solution to scale effects has received attention. The scale conversion model can characterize the numerical or physical relationship of RS land surface parameter images at different resolutions (scales) and can quantitatively describe scale effects. This paper will also focus on the research progress of the spatial down-scaling and the spatiotemporal scaling.

2. Downscaling of NDVI based on fractal IFS

2.1 Review of downscaling of RS land surface parameters

Liang [1] has reviewed several current downscaling methods, including linear decomposition methods and nonlinear statistical decomposition methods, methods for generating continuous regions, normalized difference vegetation index (NDVI) time series decomposition, multi-resolution data fusion, the statistical downscaling method of global climate model products (GCM), etc. Further, Gao et al. [5], Zhu et al. [6], and Huang et al. [7, 8] have done systematic and effective work in the spatiotemporal fusion downscaling of land surface reflectance, which has become a research hot topic. The spectral-spatial feature fusion by Wang et al. [9–12] and Shi and Wang [13] also achieved good results for subpixel mapping. These studies, however, scarcely considered the scale conversion process from the perspective of dynamics, which studies of surface parameter downscaling based on the fractal iterated function system (IFS) have paid attention to.

As a fractal branch of mathematics, because of its complete and rigorous theoretical system, it can systematically study the performance, nature, and causes of multi-scale characteristics of natural phenomena. In the fractal geometry theory system, in addition to the familiar fractal phenomenon description and fractal measurement, the internal causes or dynamic processes of mathematical fractals (interaction, feedback, and iteration, represented by IFS-iteration function system) and the physical causes of statistical fractals (such as critical or abrupt changes) are also important research contents of fractal geometry, and fractal geometry has become a part of nonlinear dynamics research [14]. Although the current research on fractal dynamics has just started, there are still many problems waiting to be solved, but its potential value and significance in dynamics research cannot be denied.

In quantitative remote sensing research, fractal methods are mostly used in the mapping of surface morphology (spatial structure) such as active radar imagery and snow and ocean imagery [15], but it also has important applications in scale conversion research and is further deepened and expanded. The use of fractals for surface parameter scale conversion modeling usually contains two important research components:

1. The performance of fractal features, that is, fractal metrics, and also the fractal dimension of the research object. For example, Zhang et al. [16, 17] used the information dimension method to describe the fractal dimension of leaf area index (LAI) scale conversion. Luan et al. [18, 19] and Wu et al. [20] used the similar dimension method to measure the fractal dimension of NDVI and LAI scale-up conversion, respectively.

2. The intrinsic nature of the fractal phenomenon, that is, the dynamics produced, which is the combined effect of multifactor surface effects.

The mathematical basis of fractal generation is IFS. Kim and Barros [21] first constructed the r function from the dynamic factors (soil sediment content, vegetation water content) of soil moisture scale conversion and then established the IFS to describe the soil moisture downscaling, and the conversion effect was good. The model can describe the dynamic process of soil moisture scale conversion, which has physical significance and demonstrates the advantages of downscaling surface parameters based on fractal IFS. In general, there is currently little research into the causes of fractal dynamics. In mathematics, the fractal IFS is a continuously iterative calculation based on the whole research object [14], and the RS land surface parameter image is created in units of local pixels. This ensures that the mathematical IFS vertical conversion factor (*r* function) is usually constant [21], while the vertical conversion factor of RS land surface parameters (such as soil moisture) is based on the physical elements of each pixel (such as sandy soil). The amount of space and the vegetation water content varies dynamically and temporally [21]. This is why the IFS function can describe the scale switching dynamics of surface parameters and why the model has certain physical meanings. The vertical conversion factor is used to describe the interscale conversion of surface parameter values and is the key to determining the IFS function. Different surface parameters have different values due to the spatial distribution and scale conversion factors (or dynamic factors), and the vertical conversion factor (*r* function) contains different types of variables and function forms. How to determine the *r* function is the difficulty in determining the IFS function, which is also an important reason why the latter is less frequently applied in descriptions of quantitative RS land surface parameter scale conversion. Therefore, the NDVI downscaling model based on the fractal IFS function can be considered to describe the dynamic process of scale conversion. This research covers a wide area and is of great significance. The following is a description of a preliminary implementation [22].

2.2 Methodology

How does one build an NDVI downscaling model based on the fractal IFS function? The following points need to be considered: first, how to identify the sensitive factors affecting the spatial distribution and scale effect of NDVI for NDVI; second, how to use this sensitive factor to establish the vertical scale conversion factor *r* function in the IFS and then determine the IFS function to achieve NDVI downscaling; and finally, how to evaluate the downscaled conversion results. The solution incorporating these considerations is described below.

2.2.1 Identify sensitive factors

According to the above description, water body is an important parameter affecting the spatial distribution and scale effect of NDVI; thus it can be determined that the pixel water parameter is one of the important dynamic factors of NDVI scale conversion. In addition, Wen et al. [23] gave a method for albedo conversion from small-scale to large-scale images and used the pixel topographical influencing factors to correct the converted results, which demonstrated that the method was effective for albedo scale conversion of rugged terrain. Considering the close relationship between the surface reflectivity and the surface albedo, and that the surface reflectance is the basic parameter for calculating NDVI, the topographic factor parameter can be determined as one of the important kinetic factors for NDVI scale conversion. Therefore, the important dynamic factors in NDVI spatial distribution and scale conversion are determined to be the pixel water parameters and topographic factors.

2.2.2 Determine the vertical conversion factor r function and establish the IFS function

Referring to Kim [21], IFS formula (1), horizontal transformation formula (2), and vertical transformation formula (3) for large-scale surface parameter pixel downscaling are obtained as follows. The IFS formula is calculated by pixel-by-pixel sliding. Get the full image downscaling results:

$$IFS^{i,j}\big|_{n,m}\left(x^i,y^j,s^{ij}\right) = \left(p_n\left(x^i\right),q_m\left(y^j\right),I_{n,m}\left(x^i,y^j,s^{ij}\right)\right), \tag{1}$$

$$\begin{cases} p_n\left(x^i\right) = x^i_{n-1} + \alpha\left(x^i - x^i_0\right) \\ q_m\left(y^i\right) = y^j_{m-1} + \alpha\left(y^j - y^j_0\right) \end{cases}, \tag{2}$$

$$I_{n,m}\left(x^i,y^j,s^{ij}\right) = \left(e_{n,m}x^i + f_{n,m}y^j + g_{n,m}x^iy^j + r_1\left(x^i,y^j\right)s^{ij} + k_{n,m}\right) \times r_2\left(x^i,y^j\right), \tag{3}$$

where $IFS^{i,j}\big|_{n,m}\left(x^i,y^j,s^{ij}\right)$ represents the surface parameter of the pixel at the (i,j) location when the large-scale pixel of the surface parameter is downscaled to the small-scale image of the $n \times m$ dimension; x^i, y^j, and s^{ij} correspond, respectively, to the x-direction coordinate $p_n\left(x^i\right)$, the y-direction coordinate $q_m\left(y^j\right)$, and the surface parameter values $I_{n,m}\left(x^i,y^j,s^{ij}\right)$ of the three-dimensional data of the pixel; x^i_{n-1} and x^i_0 represent, respectively, the x-direction starting coordinate of the (i,j) pixel in the $n \times m$ dimensional small-scale image and the x-direction starting coordinate of the large-scale pixel; α represents the downscaling ratio (small-scale/large-scale, which is less than or equal to 1); $e_{n,m}, f_{n,m}, g_{n,m}$, and $k_{n,m}$ are, respectively, functions of the x and y coordinates of the lower left corner and upper right corner of the large-scale pixel, the downscaled surface parameter data, and the vertical scale conversion surface function; and $r_1\left(x^i,y^j\right)$ and $r_2\left(x^i,y^j\right)$ represent, respectively, the two different vertical conversion factors in the vertical scale conversion surface function. Reference [21] should be consulted for the parameters or factors not represented in the formula, which will not be explained here. Generally, the $p_n\left(x^i\right)$ and $q_m\left(y^j\right)$ coordinates of the (i,j) pixel are obtained by dividing the large-scale pixel equally into $1/\alpha$ parts, and the $I_{n,m}\left(x^i,y^j,s^{ij}\right)$ calculation is the key. In formula (3), $r_2\left(x^i,y^j\right)$ is the same as the $r_1\left(x^i,y^j\right)$ function, but their argument coefficients are different.

For NDVI, $g_{n,\mathrm{m}}$, $e_{n,\mathrm{m}}$, $f_{n,\mathrm{m}}$, and $k_{n,\mathrm{m}}$ represent the functions of the (n,m) pixel of the downscaled NDVI image. Based on the special downscaled NDVI 3D values of the four corner pixels, $g_{n,\mathrm{m}}$, $e_{n,\mathrm{m}}$, $f_{n,\mathrm{m}}$, and $k_{n,\mathrm{m}}$ can be calculated as formulas (4)–(11):

$$g_{n,\mathrm{m}} = \frac{s_{n-1,m-1} - s_{n-1,m} - s_{n,m-1} + s_{n,m} - R^g_1}{x_0y_0 - x_Ny_0 - x_0y_M + x_Ny_M}, \tag{4}$$

$$e_{n,\mathrm{m}} = \frac{s_{n-1,m-1} - s_{n,m-1} - g_{n,\mathrm{m}}\left(x_0y_0 - x_Ny_0\right) - R^e_1}{x_0 - x_N}, \tag{5}$$

$$f_{n,\mathrm{m}} = \frac{s_{n-1,m-1} - s_{n-1,m} - g_{n,\mathrm{m}}\left(x_0y_0 - x_0y_M\right) - R^f_1}{y_0 - y_M}, \tag{6}$$

$$k_{n,\mathrm{m}} = s_{n,m} - e_{n,\mathrm{m}}x_N - f_{n,\mathrm{m}}y_M - g_{n,\mathrm{m}}x_Ny_M - R^k_1. \tag{7}$$

Furthermore,

$$R_1^g = r_1\left(x_{n-1}^i, y_{m-1}^j\right)s_{0,0}^{ij} - r_1\left(x_n^i, y_{m-1}^j\right)s_{N,0}^{ij} - r_1\left(x_{n-1}^i, y_m^j\right)s_{0,M}^{ij} + r_1\left(x_n^i, y_m^j\right)s_{N,M}^{ij}, \tag{8}$$

$$R_1^e = r_1\left(x_{n-1}^i, y_{m-1}^j\right)s_{0,0}^{ij} - r_1\left(x_n^i, y_{m-1}^j\right)s_{N,0}^{ij}, \tag{9}$$

$$R_1^f = r_1\left(x_{n-1}^i, y_{m-1}^j\right)s_{0,0}^{ij} - r_1\left(x_{n-1}^i, y_m^j\right)s_{0,M}^{ij}, \tag{10}$$

$$R_1^k = r_1\left(x_n^i, y_m^j\right)s_{N,M}^{ij}. \tag{11}$$

Therefore, the calculation of the $r_1(x^i, y^j)$ function is significant, and $r_1(n, m)$ $(0 \le r_1 \le 1)$ is used to adjust the NDVI surface roughness. The following treatment focuses on establishing the vertical transformation formula for NDVI, that is, the determination of the r function (containing $r_1(x^i, y^j)$ and $r_2(x^i, y^j)$).

Based on the above sensitivity factors, a vertical conversion factor r function can be constructed:

$$r = \gamma \times S_{\text{water}} + \beta \times s + \delta, \tag{12}$$

where S_{water} represents the pixel water parameter; s represents the topographic information, taking into account the magnitude of the r function; the normalized difference water index (NDWI) and slope (calculated from the digital elevation model (DEM) image) represent, respectively, the water body effect and the topographic influence in the pixel; γ and β are the coefficients of the two parameters, respectively; and δ represents the adjustment constant. Two different orders of magnitude of r are calculated as follows:

$$r_1 = \gamma_1 \times S_{\text{water}} + \beta_1 \times s + \delta_1, \tag{13}$$

$$r_2 = \gamma_2 \times S_{\text{water}} + \beta_2 \times s + \delta_2. \tag{14}$$

For NDVI, the γ, β, and δ coefficients can be calculated by linear regression between the high-resolution NDVI image and its NDWI/slope images.

Following construction of the r function, formulas (1)–(3) can be solved in combination with other known conditions, and NDVI downscaling can be achieved.

2.2.3 Evaluation of downscaling results

In order to obtain more accurate downscaling results, if the resolution of the low resolution image is too different from the resolution of the target resolution image (such as downscaling from 250 m MODIS NDVI to 30 m NDVI), a hierarchical downscaling method will be adopted. First, the low-resolution surface parameter image is downscaled to an intermediate resolution image, and then the intermediate resolution image is further downscaled to the target resolution image, which can largely guarantee the accuracy of the result.

Referring to the study by Kim and Barros [21], the accuracy of the downscaled results can be evaluated using statistical indicators such as the maximum, minimum, variance, and standard deviation (compared to high-resolution NDVI images). Moreover, the histograms of the downscaled NDVI and true NDVI images were drawn and compared, and their correlation coefficient was calculated. With those indexes, the accuracy of the downscaled images and methodology could be validated.

2.3 Experiment and result analysis

2.3.1 Experiment

As the best indicator of the status of vegetation growth and vegetation coverage, the normalized difference vegetation index is widely used in the study of environmental (climate) changes, crop yield estimation, and other fields. Among existing vegetation index products, the moderate resolution imaging spectroradiometer

Figure 1.
30 m OLI NDVI image.

Figure 2.
240 m MOD13 Q1 image.

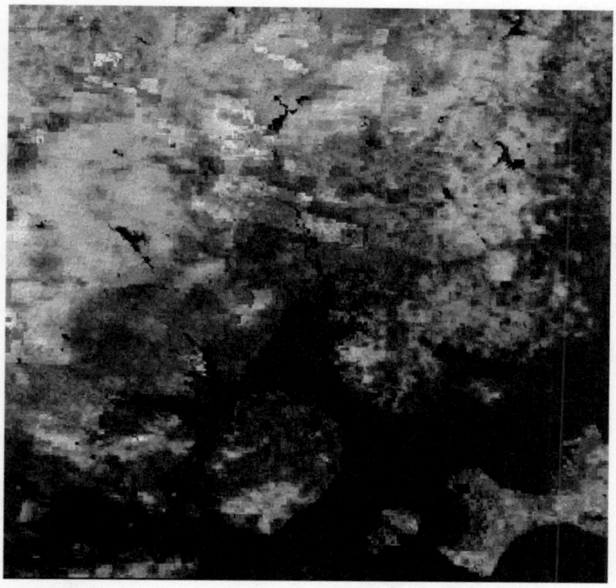

Figure 3.
30 m downscaled image of MOD13 Q.

(MODIS) vegetation index products are highly valued for their ease of use, ready availability, global coverage, and continuous phase. They have been widely used in studies of forest fires [24, 25], grassland vegetation growth [26, 27], drought [28, 29], land desertification [30], and other studies involving ecological environment monitoring. The maximum spatial resolution of MODIS vegetation index products, however, is only 250 m. The validation of this remote sensing land surface parameter is an important issue that cannot be avoided [31–33] and needs to be carried out by means of scale conversion. The most representative MODIS NDVI product, namely, MOD13 Q1, will be studied in this paper, which will also focus on establishing a downscaled model of NDVI and validating the MOD13 Q1 product based on it.

This is the experiment. A Landsat8 OLI NDVI image (**Figure 1**) was utilized to validate a MODIS NDVI image (MOD13 Q1, **Figure 2**) with nearest imaging time in Xiamen, China. Based on the downscaling formulas in Section 2.2, the MOD13 Q1 image of Xiamen was directly downscaled by ⅛ multiples, and the 30 m downscaled NDVI was obtained as **Figure 3**. The histograms of the original and processed NDVI images are drawn as **Figure 4**, and the statistics and correlation coefficients of the NDVI images are presented in **Table 1**. Based on these data, the downscaled results were evaluated and the MOD13 Q1 image was validated.

2.3.2 Result analysis

By analyzing **Figures 1–4** and **Table 1**, it is found that:

1. Compared with the real 30 m OLI NDVI image, the 30 m downscaled MOD13 Q1 image has smaller differences in maximum value, minimum value, mean value, and variance. The correlation coefficient between the two images is 0.93, which is highly correlated. The overall quality of the NDVI image obtained by downscaling the MOD13 Q1 image is considered to be good, indicating that the overall quality of MOD13 Q1 is good.

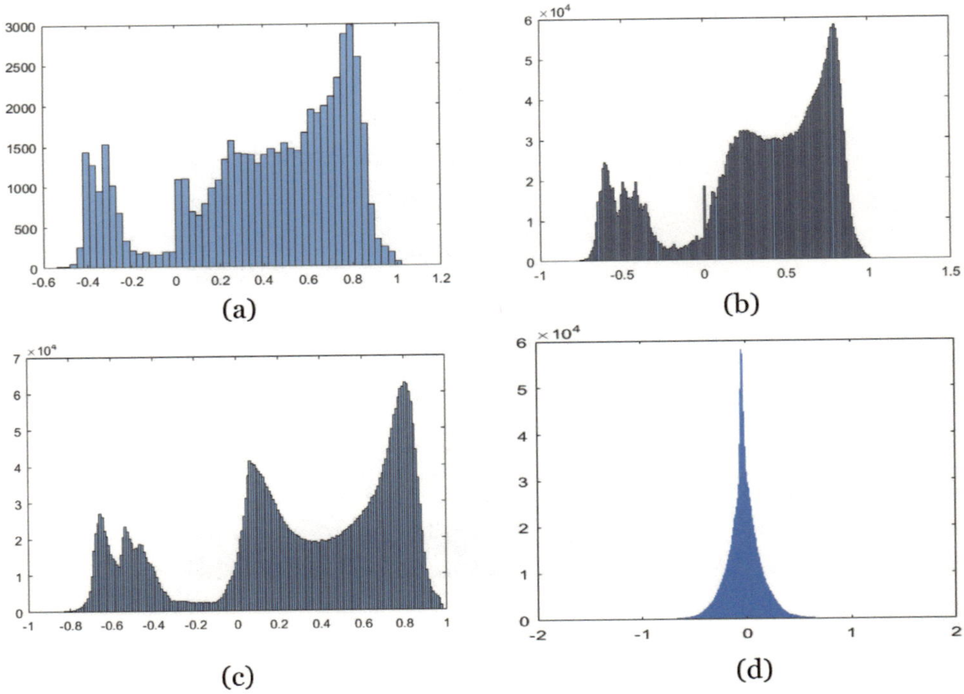

Figure 4.
Histograms of original and processed NDVI images. (a) MOD13 Q1, (b) 30 m downscaled MOD13 Q1, (c) OLI NDVI, (d) difference image between **Figures 3** *and* **1** *(***Figure 3**–**Figure 1***).*

NDVI images	Maximum	Minimum	Mean	Variance	Correlation coefficient
MOD13 Q1	0.999100	−0.561500	0.397400	0.001100	—
30 m downscaled MOD13 Q1	0.999950	−0.999793	0.326412	0.206495	0.937449
OLI NDVI	0.984436	−0.961210	0.299418	0.227313	
Difference image (**Figure 3**–**Figure 1**)	1.672050	−1.088930	−0.029594	0.027916	—

Note: The correlation coefficient is the one between the downscaled MOD13 Q1 and the OLI NDVI.

Table 1.
Statistics of original and processed NDVI images.

Comparing **Figure 4(a)** and **(b)**, there is a certain similarity between the distribution patterns of the two images, which indicates that the downscaled image retains the spatial distribution structure of the original image to a good degree, which proves to some extent that the original MOD13 Q1 image is of good quality. In addition, comparing **Figure 4(b)** and **(c)**, it is found that the downscaled NDVI image has a higher proportion in the vicinity of the zero value (mainly artificial features) than the real image. In the range of 0.2–0.6, the difference is greater. The downscaled image generally has a higher proportion in this range of values, and the histogram is smoother, indicating that the image recognition of the NDVI difference is not high. Referring to the correlation between **Figure 4(b)** and **(a)**, it is known that MOD13 Q1 also has these problems within the abovementioned range of values. Analysis of the original MOD13 Q1 image shows that it is a 16-day NDVI composite product, and each pixel takes the maximum value of NDVI within 16 days as the result of the product release. Therefore, the histogram is reasonable to a certain degree in the larger value area. At the same time, the histogram distribution of the difference image indicates that the pixel values are distributed in the range $[-1, 1]$ and the distribution pattern is low on both sides and high in the middle (approximately a value of 0), which also indicates that the downscaled image and the real image are highly consistent.

2. Further, the analysis of **Table 1** shows that the maximum value of the difference image exceeds the range of $[-1, 1]$. This may be due to a certain error which is caused by the MOD13 Q1 and OLI image during preprocessing process (atmospheric correction, geometric correction, etc.), which causes a large difference in pixel values between the MOD13 Q1 downscale image and the OLI NDVI image. However, the analysis of **Table 1** shows that the mean and variance of the difference image are small, so the above abnormal situation only occupies a small space and does not affect the overall evaluation conclusion.

According to the above analysis, the overall quality of the MOD13 Q1 downscaled image is good, indicating that the overall quality of MOD13 Q1 is good. In the NDVI range of values from 0.2 to 0.6, MOD13 Q1 is overestimated, and its discrimination ability of NDVI difference is low, which should be taken into account in practical applications.

2.4 Discussion

Based on the fractal iterated function system, downscaling models of remote sensing land surface parameters can be established. The models can then be merged with more ancillary data, which relate to the scale effects of land surface parameters. Therefore, the models are of benefit for obtaining accurate downscaled results.

In summary, although the breadth and depth of the fractal IFS application in establishing RS land surface parameters downscaling models is still insufficient, the inherent physical meaning and the advantages of the dynamic process expression of this method confer great potential on it, which needs further investigation. It is expected to become a new universal method for quantitative downscaling of RS land surface parameters and lead to the discovery of new research methods.

3. Establishing spatiotemporal scale conversion models of RS land surface parameters based on multi-fractal theory and method

3.1 Review of establishing spatiotemporal scale conversion models of RS land surface parameters

The phase is an important feature of RS images. When the phase changes, the spectrum of the objects in the image changes accordingly. Then, the parameters calculated based on the spectral information will also change, such as surface reflectivity, NDVI, and so on. The temporal response of RS land surface parameters will be further reflected in the variation of its spatial scale conversion model, i.e., the phase characteristics of spatial scale effects.

In order to quantitatively characterize the phase characteristics of spatial scale effects, that is, to establish a spatiotemporal scale conversion model (also called a spatiotemporal scaling fusion model), scholars combined the advantages of higher temporal-resolution feature of low spatial-resolution images and higher spatial-resolution feature of medium spatial-resolution images, and carried out a series of studies on spatiotemporal fusion of remotely sensed surface parameters such as surface reflectance [34], land surface temperature [35, 36], vegetation indexes [37], leaf area index [38], and so on. And then Huang et al. [39] reviewed this and presented the systematic achievements in theory and application. From the theoretical basis of spatiotemporal scale conversion fusion (the spatial scale consistency of the time-phase variation model and the time consistency of the spatial downscaling model) to the type division of the spatiotemporal scale conversion fusion algorithms (the algorithms based on features' components, the algorithms based on surface spatial information, the algorithms based on features' temporal change, and the combination algorithms of the ones above), and then to the key problems and challenges encountered in existing research (the imaging geometry and radiation characteristics, differences between multi-source RS images, the complexity of subpixel unmixing models, the complexity of features' temporal change models, etc.), and the possible development trend in future (improvement in the versatility and robustness of the algorithms), he made a detailed and in-depth explanation, so that we have a more comprehensive understanding of the development of spatiotemporal scale conversion fusion research. In fact, in addition to this method, the multi-fractal method has important potential to solve the above problems [21, 40]. The following is an example of NDVI analysis and how to establish a spatiotemporal scale conversion model (or spatiotemporal scaling fusion model) based on multi-fractal theory and method.

3.2 Spatiotemporal scale conversion models of NDVI based on multi-fractal theory and method

As the best indicator of vegetation growth status and vegetation coverage, NDVI has typical phenological characteristics. This means that in the same area where the surface cover type is unchanged, the physiological characteristics and external forms of the plant can change significantly in different growth stages, and this change will be directly reflected in the changes in image spectrum and NDVI of the surface. Furthermore, the NDVI spatial scale conversion model based on RS images of different growth periods (i.e., different phases) will also change. How to effectively reflect the influence of the phase characteristics of RS images on the construction of this model and then construct a more universal NDVI spatial scale conversion model that can be integrated with surface phenological features, namely, NDVI spatiotemporal scale conversion model? This issue has important research value. Kim and Barros [21] proposed the idea of multi-fractal method for multi-temporal remote sensing soil moisture spatial down-scaling model to describe the phase characteristics of soil moisture spatial down-scaling, but did not do specific research.

Referring to the existing knowledge, the specific method of establishing the NDVI spatiotemporal scale conversion model is given here: first, analyze the surface condition of the study area, determine the type of the main cover of the study area, and based on its phenological knowledge, select enough low and medium-high spatial resolution images finely corresponding to important "nodes" of vegetation throughout the growing season; secondly, the NDVI spatial downscaling models for different growing stages "nodes" are constructed based on the down-scaling methods such as fractal IFS; third, according to multi-fractal theory and method, using the time phase as a factor in the fractal dimension calculation method, the models corresponding to each growth stages are "fused" to obtain a unified and full growth period NDVI scale conversion model (i.e., NDVI spatiotemporal scale conversion model). At this time, the time phase (i.e., different growth stages) has been embodied as a parameter in the model. This model is more universal than the downscaling model based on the single phase image. To obtain a medium to high spatial resolution NDVI image of a certain phase during vegetation growth, the corresponding phase and the low spatial resolution NDVI image of the phase are brought into the model calculation. Of course, this method requires the research object to have a more significant phase or time periodicity, and the established spatiotemporal scale conversion model is more accurate.

Besides, there is another method of multi-fractal modeling of NDVI spatiotemporal scaling. The implementation idea is similar with Section 2.2, while the r function changes. The r functional parameters may need to be recalibrated when the spatial distribution of vegetation cover changes obviously with time (e.g., sowing stage, heading stage, maturity stage, etc.). Therefore, r function will be merged with temporal parameters of NDVI distribution, such as LAI. And the multi-fractal model of NDVI spatiotemporal scaling should be a function of NDVI to capture temporal changes in relation to ancillary data such as LAI.

Although the multi-fractal theory and method has advantages in constructing a spatiotemporal scale conversion model of RS land surface parameters, the theory and implementation of this method are more complicated, and few research cases are currently seen. However, this method is expected to become a new method for the construction of spatiotemporal scale conversion model of RS land surface parameters, which is worthy of further study.

4. Conclusions

Taking normalized difference vegetation index (NDVI) as an example, the establishment of scaling models based on fractal theory was described and analyzed in the paper. It was concluded that fractal iterated function system was an effective methodology to establish downscaling models for remote sensing land surface parameters such as NDVI and multi-fractal modeling may be a novel methodology to establish spatiotemporal scale conversion models for land surface parameters such as NDVI in the future.

Acknowledgements

This work was supported by the National Natural Science Foundation of China "Coupling of NDVI's up-scaling and down-scaling fusing with ground objects classification" (No. 41601350), the Natural Science Foundation of Fujian Province, China "Research on NDVI's scaling fusing with ground objects classification" (No. 2017J05069), and the "Scientific Research Climbing Plan" Project from Xiamen University of Technology "Spatial Distribution Estimation and Dynamic Monitoring of Soil Organic Matter Based on Multi-source and Heterogeneous Data" (No. XPDKT19010).

Author details

Haijun Luan[1,2]

1 College of Computer and Information Engineering, Xiamen University of Technology, Xiamen, China

2 Big Data Institute of Digital Natural Disaster Monitoring in Fujian, Xiamen University of Technology, Xiamen, China

*Address all correspondence to: luanhaijun@xmut.edu.cn

References

[1] Liang S. Quantitative Remote Sensing. Translated by Fan WJ. Beijing: Science Press; 2009. pp. 180-183. (in Chinese)

[2] Mohanty BP. Soil hydraulic property estimation using remote sensing: A review. Vadose Zone Journal. 2013;**12**(4). DOI: 10.2136/vzj2013.06.0100. Available from: https://dl.sciencesocieties.org/publications/vzj/abstracts/12/4/vzj2013.06.0100

[3] Li XW, Wang YT. Prospects on future developments of quantitative remote sensing. Acta Geographica Sinica. 2013;**68**(9):1163-1169. (in Chinese). DOI: 10.11821/dlxb201309001

[4] Luan HJ, Tian QJ, Yu T, et al. Review of up-scaling of quantitative remote sensing. Advances in Earth Science. 2013;**28**(6):657-664. (in Chinese) 10.11867/j.issn.1001-8166.2013.06.0657

[5] Gao F, Masek J, Schwaller M, et al. On the blending of the Landsat and MODIS surface reflectance: Predicting daily Landsat surface reflectance. IEEE Transactions on Geoscience and Remote Sensing. 2006;**44**(8):2207-2218. DOI: 10.1109/TGRS.2006.872081

[6] Zhu XL, Chen JM, Gao F, et al. An enhanced spatial and temporal adaptive reflectance fusion model for complex heterogeneous regions. Remote Sensing of Environment. 2010;**114**(11): 2610-2623. DOI: 10.1016/j.rse.2010.05.032

[7] Huang B, Zhang HK, Song HH, et al. Unified fusion of remote sensing imagery: Generating simultaneously high-resolution synthetic spatial-temporal spectral earth observations. Remote Sensing Letters. 2013;**4**:561-569. DOI: 10.1080/2150704X.2013.769283

[8] Huang B, Zhang HK. Spatio-temporal reflectance fusion via unmixing: Accounting for both phenological and land-cover changes. International Journal of Remote Sensing. 2014;**35**(16):6213-6233. DOI: 10.1080/01431161.2014.951097

[9] Wang Q, Shi W, Wang L. Allocating classes for soft-then-hard subpixel mapping algorithms in units of class. IEEE Transactions on Geoscience and Remote Sensing. 2014;**52**(5):2940-2959. DOI: 10.1109/TGRS.2013.2267802

[10] Wang Q, Shi W, Atkinson PM, et al. Downscaling MODIS images with area-to-point regression kriging. Remote Sensing of Environment. 2015;**166**: 191-204. DOI: 10.1016/j.rse.2015.06.003

[11] Wang Q, Atkinson PM, Shi W. Indicator cokriging-based subpixel mapping without prior spatial structure information. IEEE Transactions on Geoscience and Remote Sensing. 2015; **53**(1):309-323. DOI: 10.1109/TGRS.2014.2321834

[12] Wang Q, Atkinson PM, Shi W. Fast sub-pixel mapping algorithms for sub-pixel resolution change detection. IEEE Transactions on Geoscience and Remote Sensing. 2015;**53**(4):1692-1706. DOI: 10.1109/TGRS.2014.2346535

[13] Shi W, Wang Q. Soft-then-hard sub-pixel mapping with multiple shifted images. International Journal of Remote Sensing. 2015;**36**(5):1329-1348. DOI: 10.1080/01431161.2015.1009650

[14] Chen Y, Chen L. Fractal Geometry. 2nd ed. Beijing: Earthquake Press; 2005. pp. 49-51, 95-98. (in Chinese)

[15] Riccio D, Ruello G. Synthesis of fractal surfaces for remote-sensing applications. IEEE Transactions on Geoscience and Remote Sensing. 2015; **53**(7):3803-3814. DOI: 10.1109/TGRS.2014.2384595

[16] Zhang RH, Tian J, Li ZL, et al. Spatial scaling and information fractal dimension of surface parameters used in quantitative remote sensing. International Journal of Remote Sensing. 2008;**29**:5145-5159. DOI: 10.1080/01431160802036581

[17] Zhang RH, Tian J, Li ZL, et al. Principles and methods for the validation of quantitative remote sensing products. Science China Earth Sciences. 2010;**53**:741-751. DOI: 10.1007/s11430-010-0021-3

[18] Luan HJ, Tian QJ, Gu XF, et al. Establishing continuous scaling of NDVI based on fractal theory and GEOEYE-1 image. Journal of Infrared and Millimeter Waves. 2013;**32**(6):538-544, 549. (in Chinese). DOI: 10.3724/SP.J.1010.2013.00538

[19] Luan HJ, Tian QJ, Yu T, et al. Establishing continuous spatial scaling model of NDVI on fractal theory and five-index estimation system. Journal of Remote Sensing. 2015;**19**(1):116-125. (in Chinese). DOI: 1007-4619(2015)01-0116-10

[20] Wu L, Qin Q, Liu X, et al. Spatial up-scaling correction for leaf area index based on the fractal theory. Remote Sensing. 2016;**8**(3):197. DOI: 10.3390/rs8030197

[21] Kim G, Barros AP. Downscaling of remotely sensed soil moisture with a modified fractal interpolation method using contraction mapping and ancillary data. Remote Sensing of Environment. 2002;**83**:400-413. DOI: 10.1016/S0034-4257(02)00044-5

[22] Luan HJ, Zhang M, Wan YY, et al. Establishing the downscaling model of NDVI based on the iterated function system. In: Lecture Notes in Electrical Engineering. In press

[23] Wen JG, Liu Q, Liu QH, et al. Scale effect and scale correction of land-surface albedo in rugged terrain. International Journal of Remote Sensing. 2009;**30**(20):5397-5420. DOI: 10.1080/01431160903130903

[24] Huesca M, Litago J, Palaciosorueta A, et al. Assessment of forest fire seasonality using MODIS fire potential: A time series approach. Agricultural and Forest Meteorology. 2009;**149**(11):1946-1955. DOI: 10.1016/j.agrformet.2009.06.022

[25] Leon JR, Van Leeuwen WJ, Casady GM, et al. Using MODIS-NDVI for the modeling of post-wildfire vegetation response as a function of environmental conditions and pre-fire restoration treatments. Remote Sensing. 2012;**4**(3):598-621. DOI: 10.3390/rs4030598

[26] Zhang B, Zhang L, Xie D, et al. Application of synthetic NDVI time series blended from Landsat and MODIS data for grassland biomass estimation. Remote Sensing. 2015;**8**(1):10. DOI: 10.3390/rs8010010

[27] Nestola E, Calfapietra C, Emmerton CA, et al. Monitoring grassland seasonal carbon dynamics, by integrating MODIS NDVI, proximal optical sampling, and Eddy covariance measurements. Remote Sensing. 2016;**8**(3):260. DOI: 10.3390/rs8030260

[28] Du L, Tian Q, Yu T, et al. A comprehensive drought monitoring method integrating MODIS and TRMM data. International Journal of Applied Earth Observation and Geoinformation. 2013;**23**(1):245-253. DOI: 10.1016/j.jag.2012.09.010

[29] Kim Y. Drought and elevation effects on MODIS vegetation indices in northern Arizona ecosystems. International Journal of Remote Sensing. 2013;**34**(14):4889-4899. DOI: 10.1080/2150704X.2013.781700

[30] Veron SR, Paruelo JM. Desertification alters the response of

vegetation to changes in precipitation. Journal of Applied Ecology. 2010;**47**(6): 1233-1241. DOI: 10.1111/j.1365-2664.2010.01883.x

[31] Gao X, Huete AR, Didan K, et al. Multisensor comparisons and validation of MODIS vegetation indices at the semiarid Jornada experimental range. IEEE Transactions on Geoscience and Remote Sensing. 2003;**41**(10): 2368-2381. DOI: 10.1109/TGRS.2003.813840

[32] Fensholt R, Sandholt I, Stisen S, et al. Evaluating MODIS, MERIS, and VEGETATION vegetation indices using in situ measurements in a semiarid environment. IEEE Transactions on Geoscience and Remote Sensing. 2006; **44**(7):1774-1786. DOI: 10.1109/TGRS.2006.875940

[33] Geng L, Ma M, Yu W, et al. Validation of the MODIS NDVI products in different land-use types using In situ measurements in the Heihe River basin. IEEE Geoscience and Remote Sensing Letters. 2014;**11**(9): 1649-1653. DOI: 10.1109/LGRS.2014.2314134

[34] Emelyanova IV, Mcvicar TR, Van Niel TG, et al. Assessing the accuracy of blending Landsat-MODIS surface reflectance in two landscapes with contrasting spatial and temporal dynamics: A framework for algorithm selection. Remote Sensing of Environment. 2013;**133**:193-209. DOI: 10.1016/j.rse.2013.02.007

[35] Huang B, Wang J, Song H, et al. Generating high spatiotemporal resolution land surface temperature for urban Heat Island monitoring. IEEE Geoscience and Remote Sensing Letters. 2013;**10**(5):1011-1015. DOI: 10.1109/LGRS.2012.2227930

[36] Kim J, Hogue TS. Evaluation and sensitivity testing of a coupled Landsat-MODIS downscaling method for land surface temperature and vegetation indices in semi-arid regions. Journal of Applied Remote Sensing. 2012;**6**(1): 063569-1-063569-17. DOI: 10.1117/1.JRS.6.063569

[37] Ouyang W, Hao F, Skidmore AK, et al. Integration of multi-sensor data to assess grassland dynamics in a Yellow River sub-watershed. Ecological Indicators. 2012;**18**(1):163-170. DOI: 10.1016/j.ecolind.2011.11.013

[38] Zhang HK, Chen JM, Huang B, et al. Reconstructing seasonal variation of Landsat vegetation index related to leaf area index by fusing with MODIS data. IEEE Journal of Selected Topics in Applied Earth Observations and Remote Sensing. 2014;**7**(3):950-960. DOI: 10.1109/JSTARS.2013.2284528

[39] Huang B, Zhao YQ. Research status and Prospect of spatiotemporal fusion of multi-source satellite remote sensign imagery. Acta Geodaetica et Cartographica Sinica. 2017;**46**(10): 1492-1499. (in Chinese). DOI: 10.11947/j.AGCS.2017.20170376

[40] Chen JM. Spatial scaling of a remotely sensed surface parameter by contexture. Remote Sensing of Environment. 1999;**69**:30-42. DOI: 10.1016/S0034-4257(99)00006-1

Chapter 4

A New BEM for Modeling and Optimization of 3T Fractional Nonlinear Generalized Magneto-Thermoelastic Multi-Material ISMFGA Structures Subjected to Moving Heat Source

Mohamed Abdelsabour Fahmy

Abstract

The main purpose of this chapter, which represents one of the chapters of a fractal analysis book, is to propose a new boundary element method (BEM) formulation based on time fractional order theory of thermoelasticity for modeling and optimization of three temperature (3T) multi-material initially stressed multilayered functionally graded anisotropic (ISMFGA) structures subjected to moving heat source. Fractional order derivative considered in the current chapter has been found to be an accurate mathematical tool for solving the difficulty of our physical and numerical modeling. Furthermore, this chapter shed light on the practical application aspects of boundary element method analysis and topology optimization of fractional order thermoelastic ISMFGA structures. Numerical examples based on the multi-material topology optimization algorithm and bi-evolutionary structural optimization method (BESO) are presented to study the effects of fractional order parameter on the optimal design of thermoelastic ISMFGA structures. The numerical results are depicted graphically to show the effects of fractional order parameter on the sensitivities of total temperature, displacement components and thermal stress components. The numerical results also show the effects of fractional order parameter on the final topology of the ISMFGA structures and demonstrate the validity and accuracy of our proposed technique.

Keywords: boundary element method, modeling and optimization, time fractional order, three-temperature, nonlinear generalized thermoelasticity, initially stressed multilayered functionally graded anisotropic structures, moving heat source

1. Introduction

The fractional calculus has recently been widely used to study the theory and applications of derivatives and integrals of arbitrary non-integer order. This branch

IntechOpen

of mathematical analysis has emerged in recent years as an effective and powerful tool for the mathematical modeling of various engineering, industrial, and materials science applications [1–3]. The fractional-order operators are useful in describing the memory and hereditary properties of various materials and processes, due to their nonlocal nature. It clearly reflects from the related literature produced by leading fractional calculus journals that the primary focus of the investigation had shifted from classical integer-order models to fractional order models [4, 5]. Fractional calculus has important applications in hereditary solid mechanics, fluid dynamics, viscoelasticity, heat conduction modeling and identification, biology, food engineering, econophysics, biophysics, biochemistry, robotics and control theory, signal and image processing, electronics, electric circuits, wave propagation, nanotechnology, etc. [6–8].

Numerous mathematicians have contributed to the history of fractional calculus, where Euler mentioned interpolating between integral orders of a derivative in 1730. Then, Laplace defined a fractional derivative by means of an integral in 1812.

Lacroix introduced the first fractional order derivative which appeared in a calculus in 1819, where he expressed the nth derivative of the function $y = x^m$ as follows:

$$\frac{d^n}{dx^n} = \frac{\Gamma(m+1)}{\Gamma(m\text{-}n+1)} x^{m\text{-}n} \tag{1}$$

Liouville assumed that $\frac{d^v}{dx^v}(e^{ax}) = a^v e^{ax}$ for $v > 0$ to obtain the following fractional order derivative:

$$\frac{d^v x^{-a}}{dx^v} = (-1)^v \frac{\Gamma(a+v)}{\Gamma(a)} x^{-a-v} \tag{2}$$

Laurent has been using the Cauchy's integral formula for complex valued analytical functions to define the integration of arbitrary order $v > 0$ as follows:

$${}_cD_x^v f(x) = {}_cD_x^{m-\rho} f(x) = \frac{d^m}{dx^m}\left[\frac{1}{\Gamma(\rho)}\int_c^x (x-t)^{\rho-1} f(t)dt\right], 0 < \rho \leq 1 \tag{3}$$

where ${}_cD_x^v$ denotes differentiation of order v of the function f along the x-axis. Cauchy introduced the following fractional order derivative:

$$f_+^{(\alpha)} = \int f(\tau)\frac{(t\text{-}\tau)^{\alpha-1}}{\Gamma(-\alpha)} d\tau \tag{4}$$

Caputo introduced his fractional derivative of order $\alpha < 0$ to be defined as follows:

$$D_*^\alpha f(t) = \frac{1}{\Gamma(m\text{-}\alpha)}\int_0^t \frac{f^{(m)}(\tau)}{(t\text{-}\tau)^{\alpha+1-m}} d\tau, m-1 < \alpha < m, \alpha > 0 \tag{5}$$

Recently, research on nonlinear generalized magneto-thermoelastic problems has received wide attention due to its practical applications in various fields such as geomechanics, geophysics, petroleum and mineral prospecting, earthquake engineering, astronautics, oceanology, aeronautics, materials science, fiber-optic communication, fluid mechanics, automobile industries, aircraft, space vehicles, plasma physics, nuclear reactors, and other industrial applications. Due to computational difficulties in solving nonlinear generalized magneto-thermoelastic problems in general analytically, many numerical techniques have been developed and implemented for solving such problems [9–17]. The boundary element method (BEM) [18–31] has been recognized as an attractive alternative numerical method to domain methods [32–36] like finite difference method (FDM), finite element method (FEM), and finite volume method (FVM) in engineering applications. The superior feature of BEM over domain methods is that only the boundary of the domain needs to be discretized, which often leads to fewer elements and easier to use. This advantage of BEM over other domain methods has significant importance for modeling and optimization of thermoelastic problems which can be implemented using BEM with little cost and less input data. Nowadays, the BEM has emerged as an accurate and efficient computational technique for solving complicated inhomogeneous and non-linear problems in physical and engineering applications [37–69].

In the present chapter, we introduce a practical engineering application of fractal analysis in the field of thermoelasticity, where the thermal field is described by time fractional three-temperature radiative heat conduction equations. Fractional order derivative considered in the current chapter has high ability to remove the difficulty of our numerical modeling. A new boundary element method for modeling and optimization of 3T fractional order nonlinear generalized thermoelastic multi-material initially stressed multilayered functionally graded anisotropic (ISMFGA) structures subjected to moving heat source is investigated. Numerical results show that the fractional order parameter has a significant effect on the sensitivities of displacements, total three-temperature, and thermal stresses. Numerical examples show that the fractional order parameter has a significant effect on the final topology of ISMFGA structures. Numerical results of the proposed model confirm the validity and accuracy of the proposed technique, and numerical examples results demonstrate the validity of the BESO multi-material topology optimization method.

A brief summary of the chapter is as follows: Section 1 introduces the background and provides the readers with the necessary information to books and articles for a better understanding of fractional order problems and their applications. Section 2 describes the physical modeling of fractional order problems in three-temperature nonlinear generalized magneto-thermoelastic ISMFGA structures. Section 3 outlines the BEM implementation for modeling of 3T fractional nonlinear generalized magneto-thermoelastic problems of multi-material ISMFGA structures subjected to moving heat source. Section 4 introduces an illustration of the mechanisms of solving design sensitivities and optimization problem of the current chapter. Section 5 presents the new numerical results that describe the effects of fractional order parameter on the problem's field variations and on the final topology of multi-material ISMFGA structures.

2. Formulation of the problem

Consider a multilayered structure with n functionally graded layers in the xy-plane of a Cartesian coordinate. The x-axis is the common normal to all layers as shown in **Figure 1**. The thickness of the layer is denoted by h. The considered multilayered structure has been placed in a primary magnetic field H_0 acting in the direction of the y-axis.

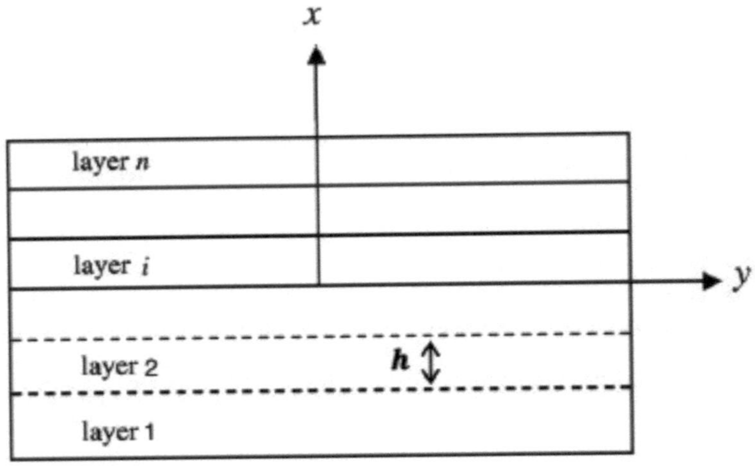

Figure 1.
Geometry of the considered problem.

According to the three-temperature theory, the governing equations of nonlinear generalized magneto-thermoelasticity in an initially stressed multilayered functionally graded anisotropic (ISMFGA) structure for the ith layer can be written in the following form:

$$\sigma_{ab,b} + \tau_{ab,b} - \Gamma_{ab} = \rho^i (x+1)^m \ddot{u}^i_a \tag{6}$$

$$\sigma_{ab} = (x+1)^m \left[C^i_{abfg} u^i_{f,g} - \beta^i_{ab}\left(T^i_\alpha - T^i_{\alpha 0} + \tau_1 \dot{T}^1_\alpha\right)\right] \tag{7}$$

$$\tau_{ab} = \mu^i (x+1)^m \left(\tilde{h}_a H_b + \tilde{h}_b H_a - \delta_{ba}\left(\tilde{h}_f H_f\right)\right) \tag{8}$$

$$\Gamma_{ab} = P^i (x+1)^m \left(\frac{\partial u^i_a}{\partial x_b} - \frac{\partial u^i_b}{\partial x_a}\right) \tag{9}$$

According to Fahmy [10], the time fractional order two-dimensional three-temperature (2D-3 T) radiative heat conduction equations in nondimensionless form can be expressed as follows:

$$D^a_\tau T^i_\alpha(r,\tau) = \xi \nabla\left[\mathbb{K}^i_\alpha \nabla T^i_\alpha(r,\tau)\right] + \xi \overline{\mathbb{W}}(r,\tau), \qquad \xi = \frac{1}{c^i_{s\alpha}\rho^i \delta_1} \tag{10}$$

where

$$\overline{\mathbb{W}}(r,\tau) = \begin{cases} -\rho^i \mathbb{W}_{eI}\left(T^i_e - T^i_I\right) - \rho^i \mathbb{W}_{er}\left(T^i_e - T^i_p\right) + \overline{\overline{\mathbb{W}}}, & \alpha = e, \delta_1 = 1 \\ \rho^i \mathbb{W}_{eI}\left(T^i_e - T^i_I\right) + \overline{\overline{\mathbb{W}}}, & \alpha = I, \delta_1 = 1 \\ \rho^i \mathbb{W}_{er}\left(T^i_e - T^i_p\right) + \overline{\overline{\mathbb{W}}}, & \alpha = p, \delta_1 = T^3_p \end{cases} \tag{11}$$

in which

$$\overline{\overline{\mathbb{W}}}(r,\tau) = -\mathbb{K}^i_\alpha \dot{T}^i_{\alpha,ab} + \beta^i_{ab} T^i_{a0} \tau_0 \ddot{u}^i_{a,b} + \rho^i c^i_{sa} \tau_0 \dddot{T}^i_\alpha - Q(x,\tau) \tag{12}$$

$$\mathbb{W}_{eI} = \rho^i \mathbb{A}_{eI} T_e^{i\left(-\frac{2}{3}\right)}, \mathbb{W}_{er} = \rho^i \mathbb{A}_{er} T_e^{i\left(-\frac{1}{2}\right)}, \mathbb{K}_\alpha = \mathbb{A}_\alpha T_\alpha^{i\left(\frac{5}{2}\right)}, \alpha = e, I, \mathbb{K}_p = \mathbb{A}_p T_p^{i(3+\mathbb{B})} \tag{13}$$

The total energy of unit mass can be described by

$$P = P_e + P_I + p_p, P_e = c_e T_e^i, P_I = c_I T_I^i, P_p = \frac{1}{4} c_p T_p^{4i} \tag{14}$$

where σ_{ab}, τ_{ab}, and u_k^i are mechanical stress tensor, Maxwell's electromagnetic stress tensor, and displacement vector in the ith layer, respectively, $c_\alpha(\alpha = c, I, p)$ are constant $T_{\alpha 0}^i$, T_α^i, C_{abfg}^i, and β_{ab}^i are, respectively, reference temperature, temperature, constant elastic moduli, and stress-temperature coefficients in the ith layer: μ^i, $\tilde{h}$, P^i, ρ^i, and $c_{s\alpha}^i$ are, respectively, magnetic permeability, perturbed magnetic field, initial stress, density, isochore specific heat coefficients in the ith layer; τ is the time; τ_0 and τ_1 are the relaxation times; $i = 1, 2, \dots, n$ represents the parameters in multilayered structure; and m is a functionally graded parameter. Also, we considered in the current study that the medium is subjected to a moving heat source of constant strength moving along x-axis with a constant velocity v. This moving heat source is assumed to have the following form:

$$Q(x, \tau) = Q_0 \delta(x - v\tau) \tag{15}$$

where, Q_0 is the heat source strength and δ is the delta function.

3. BEM numerical implementation

By using Eqs. (7)–(9), we can write (6) as

$$L_{gb} u_f^i = \rho^i \ddot{u}_a^i - \left(D_a T_\alpha^i - P^i \left(\frac{\partial u_b^i}{\partial x_a} - \frac{\partial u_a^i}{\partial x_b} \right) \right) = f_{gb} \tag{16}$$

where inertia term, temperature gradient, and initial stress terms are treated as the body forces.

In this section, we are interested in using a boundary element method for modeling the two-dimensional three-temperature radiation heat conduction equations coupled with electron, ion, and phonon temperatures.

According to finite difference scheme of Caputo at times $(f+1)\Delta t$ and $f\Delta\tau$, we obtain [1].

$$D_\tau^a T_\alpha^{i(f+1)} + D_\tau^a T_\alpha^{i(f)} \approx \sum_{j=0}^{k} W_{a,j} \left(T_\alpha^{i(f+1-j)}(r) - T_\alpha^{i(f-j)}(r) \right) \tag{17}$$

where

$$W_{a,0} = \frac{(\Delta\tau)^{-a}}{\Gamma(2-a)}, W_{a,j} = W_{a,0} \left((j+1)^{1-a} - (j-1)^{1-a} \right) \tag{18}$$

Based on Eq. (17), the fractional order heat Eq. (10) can be replaced by the following system:

$$\begin{aligned} &W_{a,0} T_\alpha^{i(f+1)}(r) - \mathbb{K}_\alpha(x) T_{\alpha,II}^{i(f+1)}(r) - \mathbb{K}_{\alpha,I}(x) T_{\alpha,I}^{i(f+1)}(r) = W_{a,0} T_\alpha^{i(f)}(r) - \mathbb{K}_\alpha(x) T_{\alpha,II}^{i(f)}(r) \\ &- \mathbb{K}_{\alpha,I}(x) T_{\alpha,J}^{i(f)}(r) - \sum_{j=1}^{f} W_{a,j} \left(T_\alpha^{i(f+1-j)}(r) - T_\alpha^{i(f-j)}(r) \right) + \overline{\mathbb{W}}_m^{i(f+1)}(x,\tau) + \overline{\mathbb{W}}_m^{i(f)}(x,\tau) \end{aligned} \tag{19}$$

where $j = 1, 2, \ldots, F, f = 0, 1, 2, \ldots, F$.

Now, according to Fahmy [10], and applying the fundamental solution which satisfies (19), the boundary integral equations corresponding to (10) without heat sources can be expressed as

$$T^i_\alpha(\xi) = \int_S \left[T^i_\alpha q^{i*} - T^{i*}_\alpha q^i\right] dC - \int_R f_{ab} T^{i*}_\alpha \, dR \tag{20}$$

Thus, the governing equations can be written in operator form as follows:

$$L_{gb} u^i_f = f_{gb}, \tag{21}$$

$$L_{ab} T^i_\alpha = f_{ab} \tag{22}$$

where the operators L_{gb}, f_{gb}, L_{ab}, and f_{ab} are as follows:

$$L_{gb} D_{abf} \frac{\partial}{\partial x_b} + D_{af} + \Lambda D_{a1f}, \quad L_{ab} = D^a_\tau \tag{23}$$

$$f_{gb} = \rho^i \ddot{u}^i_a - \left(D_a T^i_\alpha - P^i \left(\frac{\partial u^i_b}{\partial x_a} - \frac{\partial u^i_a}{\partial x_b}\right)\right) \tag{24}$$

$$f_{ab} = \frac{\mathbb{K}_\alpha}{D} T^i_\alpha \frac{\partial}{\partial \tau} \tag{25}$$

where

$$D_{abf} = C_{abfg}\varepsilon, \quad \varepsilon = \frac{\partial}{\partial x_g}, \quad D_{af} = \mu H_0^2 \left(\frac{\partial}{\partial x_a} + \delta_{a1}\Lambda\right) \frac{\partial}{\partial x_f},$$

$$D_a = -\beta^i_{ab} \left(\frac{\partial}{\partial x_b} + \delta_{b1}\Lambda + \tau_1 \left(\frac{\partial}{\partial x_b} + \Lambda\right) \frac{\partial}{\partial \tau}\right), \quad \Lambda = \frac{m}{x+1}.$$

The differential Eq. (21) can be solved using the weighted residual method (WRM) to obtain the following integral equation:

$$\int_R \left(L_{gb} u^i_f - f_{gb}\right) u^{i*}_{da} \, dR = 0 \tag{26}$$

Now, the fundamental solution u^{i*}_{df} and traction vectors t^{i*}_{da} and t^i_a can be written as follows:

$$L_{gb} u^{i*}_{df} = -\delta_{ad}\delta(x, \xi) \tag{27}$$

$$t^{i*}_{da} = C_{abfg} u^{i*}_{df,g} n_b \tag{28}$$

$$t^i_a = \frac{\bar{t}^i_a}{(x+1)^m} = \left(C_{abfg} u^i_{f,g} - \beta^i_{ab}\left(T^i_\alpha + \tau_1 \dot{T}^i_\alpha\right)\right) n_b \tag{29}$$

Using integration by parts and sifting property of the Dirac distribution for (26), then using Eqs. (27) and (29), we can write the following elastic integral representation formula:

$$u^i_d(\xi) = \int_C \left(u^{i*}_{da} t^i_a - t^{i*}_{da} u^i_a + u^{i*}_{da} \beta^i_{ab} T^i_\alpha n_b\right) dC - \int_R f_{gb} u^{i*}_{da} \, dR \tag{30}$$

The fundamental solution T^{i*} can be defined as

$$L_{ab}T^{i*}_{\alpha} = -\delta(x,\xi) \tag{31}$$

By using WRM and integration by parts, we can write (23) as follows:

$$\int_R \left(L_{ab}T^{i}_{\alpha}T^{i*}_{\alpha} - L_{ab}T^{i*}_{\alpha}T^{i}_{\alpha}\right)dR = \int_C \left(q^{i*}T^{i}_{\alpha} - q^{i}T^{i*}_{\alpha}\right)dC \tag{32}$$

where

$$q^{i} = -\mathbb{K}^{i}_{\alpha}T^{i}_{\alpha,b}n_a \tag{33}$$

$$q^{i*} = -\mathbb{K}^{i}_{\alpha}T^{i*}_{\alpha,b}n_a \tag{34}$$

By the use of sifting property, we obtain from (32) the thermal integral representation formula

$$T^{i}_{\alpha}(\xi) = \int_C \left(q^{i*}T^{i}_{\alpha} - q^{i}T^{i*}_{\alpha}\right)dC - \int_C f_{ab}T^{i*}_{\alpha}\,dR \tag{35}$$

By combining (30) and (35), we obtain

$$\begin{bmatrix} u^{i}_{d}(\xi) \\ T^{i}_{\alpha}(\xi) \end{bmatrix} = \int_C \left\{ -\begin{bmatrix} t^{i*}_{da} & -u^{i*}_{da}\beta_{ab}n_b \\ 0 & -q^{i*} \end{bmatrix}\begin{bmatrix} u^{i}_{a} \\ T^{i}_{\alpha} \end{bmatrix} + \begin{bmatrix} u^{i*}_{da} & 0 \\ 0 & -T^{i*}_{\alpha} \end{bmatrix}\begin{bmatrix} t^{i}_{a} \\ q^{i} \end{bmatrix} \right\} dC - \int_R \begin{bmatrix} u^{i*}_{da} & 0 \\ 0 & -T^{i*}_{\alpha} \end{bmatrix}\begin{bmatrix} f_{gb} \\ -f_{ab} \end{bmatrix} dR \tag{36}$$

The nonlinear generalized magneto-thermoelastic vectors can be written in contracted notation form as

$$U^{i}_{A} = \begin{cases} u^{i}_{a} & a = A = 1,2,3 \\ T^{i}_{\alpha} & A = 4 \end{cases} \tag{37}$$

$$\mathrm{T}^{i}_{\alpha A} = \begin{cases} t^{i}_{a} & a = A = 1,2,3 \\ q^{i} & A = 4 \end{cases} \tag{38}$$

$$U^{i*}_{DA} = \begin{cases} u^{i*}_{da} & d = D = 1,2,3; a = A = 1,2,3 \\ 0 & d = D = 1,2,3; A = 4 \\ 0 & D = 4; a = A = 1,2,3 \\ -T^{i*}_{\alpha} & D = 4; A = 4 \end{cases} \tag{39}$$

$$\tilde{T}^{i*}_{\alpha DA} = \begin{cases} t^{i*}_{da} & d = D = 1,2,3; a = A = 1,2,3 \\ -\tilde{u}^{i*}_{d} & d = D = 1,2,3; A = 4 \\ 0 & D = 4; a = A = 1,2,3 \\ -q^{i*} & D = 4; A = 4 \end{cases} \tag{40}$$

$$\tilde{u}^{i*}_{d} = u^{i*}_{da}\beta_{af}n_f \tag{41}$$

By using the above vectors, we can express (36) as

$$U^i_D(\xi) = \int_C \left(U^{i*}_{DA} \mathrm{T}^i_{\alpha A} - \tilde{T}^i_{\alpha DA} U^i_A \right) dC - \int_R U^{i*}_{DA} S_A dR \tag{42}$$

The source vector S_A can be divided as

$$S_A = S^0_A + S^T_A + S^u_A + S^{\dot{T}}_A + S^{\ddot{T}}_A + S^{\ddot{u}}_A \tag{43}$$

where

$$S^0_A = \begin{cases} 0 & A = 1,2,3 \\ Q_0 \delta(x - v\tau) & A = 4 \end{cases} \tag{44}$$

$$S^T_A = \omega_{AF} U^i_F \text{ with } \omega_{AF} = \begin{cases} -D_a & A = 1,2,3; F = 4 \\ \xi \nabla \left[\mathbb{K}^i_\alpha \nabla \right] & \text{otherwise} \end{cases} \tag{45}$$

$$S^u_A = \psi U^i_F \text{ with } \psi = \begin{cases} P^i \left(\dfrac{\partial}{\partial x_b} - \dfrac{\partial}{\partial x_a} \right) & A = 1,2,3; F = 1,2,3, \\ 0 & A = 4; F = 4 \end{cases} \tag{46}$$

$$S^{\dot{T}}_A = \Gamma_{AF} \dot{U}^i_F \text{ with } \Gamma_{AF} = \begin{cases} -\beta^i_{ab} \tau_1 \left(\dfrac{\partial}{\partial x_b} + \Lambda \right) \dfrac{\partial}{\partial \tau} & A = 4; F = 4 \\ -\mathbb{K}^i_\alpha & \text{otherwise} \end{cases} \tag{47}$$

$$S^{\ddot{T}}_A = \delta_{AF} \ddot{U}^i_F \text{ with } \delta_{AF} = \begin{cases} 0 & A = 4; F = 4 \\ \rho^i c^i_{s\alpha} \tau_0 & \text{otherwise} \end{cases} \tag{48}$$

$$S^{\ddot{u}}_A = \daleth \ddot{U}^i_F \text{ with } \daleth = \begin{cases} \rho^i & A = 1,2,3; F = 1,2,3, \\ \beta^i_{ab} T^i_{\alpha 0} \tau_0 & A = 4; F = 4 \end{cases} \tag{49}$$

The representation formula (36) can also be written in matrix form as follows:

$$\begin{aligned} [S_A] = & -\begin{bmatrix} 0 \\ Q_0 \delta(x - v\tau) \end{bmatrix} + \begin{bmatrix} -D_a T^i_\alpha \\ \xi \nabla \left[\mathbb{K}^i_\alpha \nabla T^i_\alpha (r, \tau) \right] \end{bmatrix} + \begin{bmatrix} P^i \left(u^i_{b,a} - u^i_{a,b} \right) \\ 0 \end{bmatrix} \\ & + \begin{bmatrix} -\beta^i_{ab} \tau_1 \left(\dfrac{\partial}{\partial x_b} + \Lambda \right) \dot{T}^i_\alpha \\ -\mathbb{K}^i_\alpha \dot{T}^i_\alpha \end{bmatrix} + \rho^i c^i_{s\alpha} \tau_0 \begin{bmatrix} 0 \\ \ddot{T}^i_\alpha \end{bmatrix} + \begin{bmatrix} \rho^i \ddot{u}^i_a \\ \beta^i_{ab} T^i_{\alpha 0} \tau_0 \ddot{u}^i{}_{f,g} \end{bmatrix} \end{aligned} \tag{50}$$

In order to convert the domain integral in (42) into the boundary, we approximate the source vector S_A by a series of known functions f^q_{AE} and unknown coefficients α^q_E as

$$S_A \approx \sum_{q=1}^{E} f^q_{AE} \alpha^q_E \tag{51}$$

Thus, the representation formula (42) can be written as follows:

$$U_D(\xi) = \int_C \left(U^{i*}_{DA} T^i_{\alpha A} - \tilde{T}^{i*}_{\alpha DA} U^i_A \right) dC - \sum_{q=1}^{N} \int_R U^{i*}_{DA} f^q_{AE} dR \alpha^q_E \tag{52}$$

By implementing the WRM to the following equations

$$L_{gb}u_{fe}^{iq} = f_{ae}^{q} \tag{53}$$

$$L_{ab}T_{\alpha}^{iq} = f_{pj}^{q} \tag{54}$$

Then the elastic and thermal representation formulae are given as follows (Fahmy [46]):

$$u_{de}^{iq}(\xi) = \int_C \left(u_{da}^{i*}t_{ae}^{iq} - t_{da}^{i*}u_{ae}^{iq}\right)dC - \int_R u_{da}^{i*}f_{ae}^{q}dR \tag{55}$$

$$T_{\alpha}^{iq}(\xi) = \int_C \left(q^{i*}T_{\alpha}^{iq} - q^{iq}T_{\alpha}^{i*}\right)dC - \int_R f^{q}T_{\alpha}^{i*}dR \tag{56}$$

The representation formulae (55) and (56) can be combined into the following single equation:

$$U_{DE}^{iq}(\xi) = \int_C \left(U_{DA}^{i*}T_{\alpha AE}^{iq} - T_{\alpha DA}^{i*}U_{AE}^{iq}\right)dC - \int_R U_{DA}^{i*}f_{AE}^{iq}dR \tag{57}$$

By substituting from Eq. (57) into Eq. (52), we obtain the following BEM coupled thermoelasticity formula:

$$\begin{aligned} U_D^i(\xi) &= \int_C \left(U_{DA}^{i*}T_{\alpha A}^{i} - \breve{T}_{\alpha DA}^{i*}U_A^i\right)dC \\ &+ \sum_{q=1}^{E}\left(U_{DE}^{iq}(\xi) + \int_C \left(T_{\alpha DA}^{i*}U_{AE}^{iq} - U_{DA}^{i*}T_{\alpha AE}^{iq}\right)dC\right)\alpha_E^q \end{aligned} \tag{58}$$

In order to compute the displacement sensitivity, Eq. (58) is differentiated with respect to ξ_l as follows:

$$\begin{aligned} \frac{\partial U_D^i(\xi)}{\partial \xi_l} &= -\int_C \left(U_{DA,l}^{i*}T_{\alpha A}^{i} - \breve{T}_{\alpha DA,l}^{i*}U_A^i\right)dC \\ &+ \sum_{q=1}^{E}\left(\frac{\partial U_{DE}^{iq}(\xi)}{\partial \xi_l} - \int_C \left(T_{\alpha DA,l}^{i*}U_{\alpha AE}^{iq} - U_{DA,l}^{i*}T_{\alpha AE}^{iq}\right)dC\right)\alpha_E^q \end{aligned} \tag{59}$$

According to the procedure of Fahmy [44], we can write (58) in the following form:

$$\breve{\zeta}U - \eta T_{\alpha} = (\zeta\breve{U} - \eta\wp)\overline{\alpha} \tag{60}$$

The generalized displacements and velocities are approximated in terms of known tensor functions f_{FD}^{q} and unknown coefficients γ_D^q and $\tilde{\gamma}_D^q$:

$$U_F^i \approx \sum_{q=1}^{N} f_{FD}^{q}(x)\gamma_D^q \tag{61}$$

where

$$f_{FD}^{q} = \begin{cases} f_{fd}^{q} & f = F = 1,2,3; d = D = 1,2,3 \\ f^{q} & F = 4; D = 4 \\ 0 & \text{otherwise} \end{cases} \tag{62}$$

Now, the gradients of the generalized displacements and velocities can also be approximated in terms of the tensor function derivatives as

$$U_{F,g}^{i} \approx \sum_{q=1}^{N} f_{FD,g}^{q}(x)\gamma_{K}^{q} \tag{63}$$

By substituting (63) into Eq. (45), we get

$$S_{A}^{T} = \sum_{q=1}^{N} S_{AF} f_{FD,g}^{q} \gamma_{D}^{q} \tag{64}$$

By applying the point collocation procedure of Gaul et al. [43] to Eqs. (51) and (61), we obtain

$$\check{S} = J\overline{\alpha}, \quad U^{i} = J'\gamma, \tag{65}$$

Similarly, applying the same point collocation procedure to Eqs. (64), (46), (47), (48), and (49) yields

$$\check{S}^{T_{\alpha}^{i}} = \mathcal{B}^{T}\gamma \tag{66}$$

$$S_{A}^{u} = \psi U^{i} \tag{67}$$

$$\check{S}^{T_{\alpha}^{i}} = \overline{\Gamma}_{AF}\dot{U}^{i} \tag{68}$$

$$\check{S}^{T_{\alpha}^{i}} = \overline{\delta}_{AF}\ddot{U}^{i} \tag{69}$$

$$\check{S}^{\ddot{u}} = \overline{\exists}\ddot{U}^{i} \tag{70}$$

where $\overline{\psi}$, $\overline{\Gamma}_{AF}$, $\overline{\delta}_{AF}$, and $\overline{\exists}$ are assembled using the submatrices $[\psi]$, $[\Gamma_{AF}]$, $[\delta_{AF}]$, and $[\exists]$, respectively.

Solving the system (65) for $\overline{\alpha}$ and γ yields

$$\overline{\alpha} = J^{-1}\check{S}, \quad \gamma = J'^{-1}U^{i} \tag{71}$$

Now, the coefficient $\overline{\alpha}$ can be written in terms of the unknown displacements U^{i}, velocities $\dot{U}^{i}$, and accelerations $\ddot{U}^{i}$ as

$$\overline{\alpha} = J^{-1}\left(\check{S}^{0} + \left(\mathcal{B}^{T}J'^{-1} + \overline{\psi}\right)U^{i} + \overline{\Gamma}_{AF}\dot{U}^{i} + \left(\overline{\exists} + \overline{\delta}_{AF}\right)\ddot{U}^{i}\right) \tag{72}$$

An implicit-implicit staggered algorithm has been implemented for use with the BEM to solve the governing equations which can now be written in a suitable form after substitution of Eq. (72) into Eq. (60) as

$$\overbrace{M}\ddot{U}^i + \overbrace{\Gamma}\dot{U}^i + \overbrace{K}U^i = \overbrace{\mathbb{Q}}^i \tag{73}$$

$$\overbrace{X}\ddot{T}^i_\alpha + \overbrace{A}\dot{T}^i_\alpha + \overbrace{B}T^i_\alpha = \overbrace{\mathbb{Z}}\ddot{U}^i + \overbrace{\mathbb{R}} \tag{74}$$

where $V = (\eta\check{\wp} - \zeta\breve{U})J^{-1}$, $\overbrace{M} = V(\exists + \overline{\delta}_{AF})$, $\overbrace{\Gamma} = V\overline{\Gamma}_{AF}$, $\overbrace{K} = -\check{\zeta} + V\left(\mathcal{B}^T J'^{-1} + \overline{\psi}\right)$, $\overbrace{\mathbb{Q}}^i = -\eta\check{T} + V\check{S}^0$, $\overbrace{X} = -\rho^i c^i_{sa}\tau_0$, $\overbrace{A} = -\mathbb{K}^i_\alpha$, $\overbrace{B} = \xi\nabla\left[\mathbb{K}^i_\alpha\,\nabla\right]$, $\overbrace{\mathbb{Z}} = \beta^i_{ab}T^i_{\alpha 0}\tau_0$, $\overbrace{\mathbb{R}} = -Q_0\delta(x - v\tau)$.

where $\ddot{U}^i, \dot{U}^i, U^i, T^i$ and $\overbrace{\mathbb{Q}}^i$ are, respectively, acceleration, velocity, displacement, temperature, and external force vectors, and $V, \overbrace{M}, \overbrace{\Gamma}, \overbrace{K}, \overbrace{A}$, and $\overbrace{B}$ are, respectively, volume, mass, damping, stiffness, capacity, and conductivity matrices.

In many applications, the coupling term $\overbrace{\mathbb{Z}}\ddot{U}^i_{n+1}$ that appear in the heat conduction equation is negligible. Therefore, it is easier to predict the temperature than the displacement.

Hence Eqs. (73) and (74) lead to the following coupled system of differential-algebraic equations (DAEs):

$$\overbrace{M}\ddot{U}^i_{n+1} + \overbrace{\Gamma}\dot{U}^i_{n+1} + \overbrace{K}U^i_{n+1} = \overbrace{\mathbb{Q}}^{ip}_{n+1} \tag{75}$$

$$\overbrace{X}\ddot{T}^i_{\alpha(n+1)} + \overbrace{A}\dot{T}^i_{\alpha(n+1)} + \overbrace{B}T^i_{\alpha(n+1)} = \overbrace{\mathbb{Z}}\ddot{U}^i_{n+1} + \overbrace{\mathbb{R}} \tag{76}$$

where $\overbrace{\mathbb{Q}}^{ip}_{n+1} = \eta T^{ip}_{\alpha(n+1)} + V\check{S}^0$ and $T^{ip}_{\alpha(n+1)}$ is the predicted temperature.
By integrating Eq. (73) and using Eq. (75), we get

$$\begin{aligned}\dot{U}^i_{n+1} &= \dot{U}^i_n + \frac{\Delta\tau}{2}\left(\ddot{U}^i_{n+1} + \ddot{U}^i_n\right)\\ &= \dot{U}^i_n + \frac{\Delta\tau}{2}\left[\ddot{U}^i_n + \overbrace{M}^{-1}\left(\overbrace{\mathbb{Q}}^{ip}_{n+1} - \overbrace{\Gamma}\dot{U}^i_{n+1} - \overbrace{K}U^i_{n+1}\right)\right]\end{aligned} \tag{77}$$

$$\begin{aligned}U^i_{n+1} &= U^i_n + \frac{\Delta\tau}{2}\left(\dot{U}^i_{n+1} + \dot{U}^i_n\right)\\ &= U^i_n + \Delta\tau\dot{U}^i_n + \frac{\Delta\tau^2}{4}\left[\ddot{U}^i_n + \overbrace{M}^{-1}\left(\overbrace{\mathbb{Q}}^{ip}_{n+1} - \overbrace{\Gamma}\dot{U}^i_{n+1} - \overbrace{K}U^i_{n+1}\right)\right]\end{aligned} \tag{78}$$

From Eq. (77) we have

$$\dot{U}^i_{n+1} = \overline{\gamma}^{-1}\left[\dot{U}^i_n + \frac{\Delta\tau}{2}\left[\ddot{U}^i_n + \overbrace{M}^{-1}\left(\overbrace{\mathbb{Q}}^{ip}_{n+1} - \overbrace{K}U^i_{n+1}\right)\right]\right] \tag{79}$$

where $\overline{\gamma} = \left(I + \frac{\Delta\tau}{2}\overbrace{M}^{-1}\overbrace{\Gamma}\right)$.

Substituting Eq. (79) into Eq. (78), we derive

$$U^i_{n+1} = U^i_n + \Delta\tau\dot{U}^i_n + \frac{\Delta\tau^2}{4}\left[\ddot{U}^i_n + \widehat{M}^{-1}\left(\widehat{\mathbb{Q}}^{ip}_{n+1} - \widehat{\Gamma}\,\overline{\gamma}^{-1}\left[\dot{U}^i_n + \frac{\Delta\tau}{2}\left[\ddot{U}^i_n + \widehat{M}^{-1}\left(\widehat{\mathbb{Q}}^{ip}_{n+1} - \widehat{K}\,U^i_{n+1}\right)\right]\right] - \widehat{K}\,U^i_{n+1}\right)\right] \quad (80)$$

Substituting $\dot{U}^i_{n+1}$ from Eq. (79) into Eq. (75), we obtain

$$\ddot{U}^i_{n+1} = \widehat{M}^{-1}\left[\widehat{\mathbb{Q}}^{ip}_{n+1} - \widehat{\Gamma}\left[\overline{\gamma}^{-1}\left[\dot{U}^i_n + \frac{\Delta\tau}{2}\left[\ddot{U}^i_n + \widehat{M}^{-1}\left(\widehat{\mathbb{Q}}^{ip}_{n+1} - \widehat{K}\,U^i_{n+1}\right)\right]\right]\right] - \widehat{K}\,U^i_{n+1}\right] \quad (81)$$

By integrating the heat Eq. (74) and using Eq. (76), we obtain

$$\begin{aligned}\dot{T}^i_{\alpha(n+1)} &= \dot{T}^i_n + \frac{\Delta\tau}{2}\left(\ddot{T}^i_{\alpha(n+1)} + \ddot{T}^i_{\alpha n}\right)\\ &= \dot{T}^i_{\alpha n} + \frac{\Delta\tau}{2}\left(\widehat{\mathrm{X}}^{-1}\left[\widehat{\mathbb{Z}}\,\ddot{U}^i_{n+1} + \widehat{\mathbb{R}} - \widehat{\mathrm{A}}\,\dot{T}^i_{\alpha(n+1)} - \widehat{\mathrm{B}}\,T^i_{\alpha(n+1)}\right] + \ddot{T}^i_{\alpha n}\right)\end{aligned} \quad (82)$$

$$\begin{aligned}T^i_{\alpha(n+1)} &= T^i_{\alpha n} + \frac{\Delta\tau}{2}\left(\dot{T}^i_{\alpha(n+1)} + \dot{T}^i_{\alpha n}\right)\\ &= T^i_{\alpha n} + \Delta\tau\dot{T}^i_{\alpha n} + \frac{\Delta\tau^2}{4}\left(\ddot{T}^i_{\alpha n} + \widehat{\mathrm{X}}^{-1}\left[\widehat{\mathbb{Z}}\,\ddot{U}^i_{n+1} + \widehat{\mathbb{R}} - \widehat{\mathrm{A}}\,\dot{T}^i_{\alpha(n+1)} - \widehat{\mathrm{B}}\,T^i_{\alpha(n+1)}\right]\right)\end{aligned} \quad (83)$$

From Eq. (82) we get

$$\dot{T}^i_{\alpha(n+1)} = \gamma^{-1}\left[\dot{T}^i_{\alpha n} + \frac{\Delta\tau}{2}\left(\widehat{\mathrm{X}}^{-1}\left[\widehat{\mathbb{Z}}\,\ddot{U}^i_{n+1} + \widehat{\mathbb{R}} - \widehat{\mathrm{B}}\,T^i_{\alpha(n+1)}\right] + \ddot{T}^i_{\alpha n}\right)\right] \quad (84)$$

where $\gamma = \left(I + \frac{1}{2}\widehat{\mathrm{A}}\,\Delta\tau\,\widehat{\mathrm{X}}^{-1}\right)$.

Substituting Eq. (84) into Eq. (83), we obtain

$$\begin{aligned}T^i_{\alpha(n+1)} = T^i_{\alpha n} + \Delta\tau\dot{T}^i_{\alpha n} + \frac{\Delta\tau^2}{4}\Bigg(\ddot{T}^i_{\alpha n} + \widehat{\mathrm{X}}^{-1}\Bigg[\widehat{\mathbb{Z}}\,\ddot{U}^i_{n+1} + \widehat{\mathbb{R}} - \widehat{\mathrm{A}}\Bigg(\gamma^{-1}\Bigg[\dot{T}^i_{\alpha n}\\ + \frac{\Delta\tau}{2}\left(\widehat{\mathrm{X}}^{-1}\left[\widehat{\mathbb{Z}}\,\ddot{U}^i_{n+1} + \widehat{\mathbb{R}} - \widehat{\mathrm{B}}\,T^i_{\alpha(n+1)}\right] + \ddot{T}^i_{\alpha n}\right)\Bigg]\Bigg) - \widehat{\mathrm{B}}\,T^i_{\alpha(n+1)}\Bigg]\Bigg)\end{aligned} \quad (85)$$

Substituting $\dot{T}^i_{n+1}$ from Eq. (84) into Eq. (76), we get

$$\begin{aligned}\ddot{T}^i_{\alpha(n+1)} = \widehat{\mathrm{X}}^{-1}\Bigg[\widehat{\mathbb{Z}}\,\ddot{U}^i_{n+1} + \widehat{\mathbb{R}} - \widehat{\mathrm{A}}\Bigg(\gamma^{-1}\Bigg[\dot{T}^i_{\alpha n} + \frac{\Delta\tau}{2}\Bigg(\widehat{\mathrm{X}}^{-1}\Bigg[\widehat{\mathbb{Z}}\,\ddot{U}^i_{n+1} + \widehat{\mathbb{R}}\\ - \widehat{\mathrm{B}}\,T^i_{\alpha(n+1)}\Bigg] + \ddot{T}^i_{\alpha(n+1)}\Bigg)\Bigg]\Bigg) - \widehat{\mathrm{B}}\,T^i_{\alpha(n+1)}\Bigg]\end{aligned} \quad (86)$$

Now, a displacement predicted staggered procedure for the solution of (80) and (85) is as follows:

The first step is to predict the propagation of the displacement wave field: $U^{ip}_{n+1} = U^i_n$. The second step is to substitute for $\dot{U}^i_{n+1}$ and $\ddot{U}^i_{n+1}$ from Eqs. (77) and (75), respectively, in Eq. (85) and solve the resulted equation for the three-temperature fields. The third step is to correct the displacement using the computed three-temperature fields for the Eq. (80). The fourth step is to compute $\dot{U}^i_{n+1}$, $\ddot{U}^i_{n+1}$, $\dot{T}^i_{\alpha(n+1)}$, and $\ddot{T}^i_{\alpha(n+1)}$ from Eqs. (79), (81), (82), and (86), respectively.

The continuity conditions for temperature, heat flux, displacement, and traction that have been considered in the current chapter can be expressed as

$$T^i_\alpha(x,z,\tau)\big|_{x=h^i} = T^{(i+1)}_\alpha(x,z,\tau)\big|_{x=h^i} \tag{87}$$

$$q^i(x,z,\tau)\big|_{x=h^i} = q^{(i+1)}(x,z,\tau)\big|_{x=h^i} \tag{88}$$

$$u^i_f(x,z,\tau)\Big|_{x=h^i} = u^{(i+1)}_f(x,z,\tau)\Big|_{x=h^i} \tag{89}$$

$$\bar{t}^i_a(x,z,\tau)\Big|_{x=h^i} = \bar{t}^{(i+1)}_a(x,z,\tau)\Big|_{x=h^i} \tag{90}$$

where n is the total number of layers, $\bar{t}_a$ are the tractions which is defined by $\bar{t}_a = \sigma_{ab}n_b$, and $i = 1, 2, \dots, n-1$.

The initial and boundary conditions of the present study are

$$u^i_f(x,z,0) = \dot{u}^i_f(x,z,0) = 0 \quad \text{for} \quad (x,z) \in R \cup C \tag{91}$$

$$u^i_f(x,z,\tau) = \Psi_f(x,z,\tau) \quad \text{for} \quad (x,z) \in C_3 \tag{92}$$

$$\bar{t}^i_a(x,z,\tau) = \Phi_f(x,z,\tau) \quad \text{for} \quad (x,z) \in C_4, \tau > 0 \tag{93}$$

$$T^i_\alpha(x,z,0) = T^i_\alpha(x,z,0) = 0 \quad \text{for} \quad (x,z) \in R \cup C \tag{94}$$

$$T^i_\alpha(x,z,\tau) = \bar{f}(x,z,\tau) \quad \text{for} \quad (x,z) \in C_1, \quad \tau > 0 \tag{95}$$

$$q^i(x,z,\tau) = \bar{h}(x,z,\tau) \quad \text{for} \quad (x,z) \in C_2, \quad \tau > 0 \tag{96}$$

where Ψ_f, Φ_f, f, and $\bar{h}$ are prescribed functions, $C = C_1 \cup C_2 = C_3 \cup C_4$, and $C_1 \cap C_2 = C_3 \cap C_4 = \emptyset$.

4. Design sensitivity and optimization

According to Fahmy [58, 60], the design sensitivities of displacements components and total 3T can be performed by implicit differentiation of (75) and (76), respectively, which describe the structural response with respect to the design variables, and then we can compute thermal stresses sensitivities.

The bi-directional evolutionary structural optimization (BESO) is the evolutionary topology optimization method that allows modification of the structure by either adding or removing material to or from the structure design. This addition or removal depends on the sensitivity analysis. Sensitivity analysis is the estimation of the response of the structure to the modification of design variables and is dependent on the calculation of derivatives [70–80].

The homogenized vector of thermal expansion coefficients α^H can be written in terms of the homogenized elasticity matrix D^H and the homogenized vector of stress-temperature coefficients β^H as follows:

$$\alpha^H = \left(D^H\right)^{-1}\beta^H \tag{97}$$

For the material design, the derivative of the homogenized vector of thermal expansion coefficients can be written as

$$\frac{\partial\alpha^H}{\partial X_{kl}^m} = \left(D^H\right)^{-1}\left(\frac{\partial\beta^H}{\partial X_{kl}^m} - \frac{\partial D^H}{\partial X_{kl}^m}\alpha^H\right) \tag{98}$$

where $\frac{\partial D^H}{\partial X_{kl}^m}$ and $\frac{\partial\beta^H}{\partial X_{kl}^m}$ for any lth material phase can be calculated using the adjoint variable method [73] as

$$\frac{\partial D^H}{\partial X_{kl}^m} = \frac{1}{|\Omega|}\int_Y (I - B^m U^m)^T \frac{\partial D^m}{\partial X_{kl}^m}(I - B^m U^m)dy \tag{99}$$

and

$$\frac{\partial\beta^H}{\partial X_{kl}^m} = \frac{1}{|Y|}\int_Y (I - B^m U^m)^T \frac{\partial D^m}{\partial X_{kl}^m}(\alpha^m - B^m\varphi^m)dy + \frac{1}{|\Omega|}\int_Y (I - B^m U^m)^T \frac{\partial\alpha^m}{\partial X_{kl}^m}dy \tag{100}$$

where $|\Omega|$ is the volume of the base cell.

5. Numerical examples, results, and discussion

In order to show the numerical results of this study, we consider a monoclinic graphite-epoxy as an anisotropic thermoelastic material which has the following physical constants [57].

The elasticity tensor is expressed as

$$C_{pjkl} = \begin{bmatrix} 430.1 & 130.4 & 18.2 & 0 & 0 & 201.3 \\ 130.4 & 116.7 & 21.0 & 0 & 0 & 70.1 \\ 18.2 & 21.0 & 73.6 & 0 & 0 & 2.4 \\ 0 & 0 & 0 & 19.8 & -8.0 & 0 \\ 0 & 0 & 0 & -8.0 & 29.1 & 0 \\ 201.3 & 70.1 & 2.4 & 0 & 0 & 147.3 \end{bmatrix} \text{GPa} \tag{101}$$

The mechanical temperature coefficient is

$$\beta_{pj} = \begin{bmatrix} 1.01 & 2.00 & 0 \\ 2.00 & 1.48 & 0 \\ 0 & 0 & 7.52 \end{bmatrix} \cdot 10^6 \frac{\text{N}}{\text{Km}^2} \tag{102}$$

The thermal conductivity tensor is

$$k_{pj} = \begin{bmatrix} 5.2 & 0 & 0 \\ 0 & 7.6 & 0 \\ 0 & 0 & 38.3 \end{bmatrix} \mathrm{W/Km} \tag{103}$$

Mass density $\rho = 7820\mathrm{kg/m^3}$ and heat capacity c = 461 J/kg K.

The proposed technique that has been utilized in the present chapter can be applicable to a wide range of three-temperature nonlinear generalized thermoelastic problems of ISMFGA structures. The main aim of this chapter was to assess the impact of fractional order parameter on the sensitivities of total three-temperature, displacement components, and thermal stress components.

Figure 2 shows the variation of the total temperature sensitivity along the x-axis. It was shown from this figure that the fraction order parameter has great effects on the total three-temperature sensitivity.

Figures 3 and **4** show the variation of the displacement components u_1 and u_2 along the x-axis for different values of fractional order parameter. It was noticed from these figures that the fractional order parameter has great effects on the displacement sensitivities.

Figures 5–7 show the variation of the thermal stress components σ_{11}, σ_{12}, and σ_{22}, respectively, along the x-axis for different values of fractional order parameter. It was noted from these figures that the fractional order parameter has great influences on the thermal stress sensitivities.

Since there are no available results for the three-temperature thermoelastic problems, except for Fahmy's research [10–14]. For comparison purposes with the special cases of other methods treated by other authors, we only considered one-dimensional numerical results of the considered problem. In the special case under consideration,

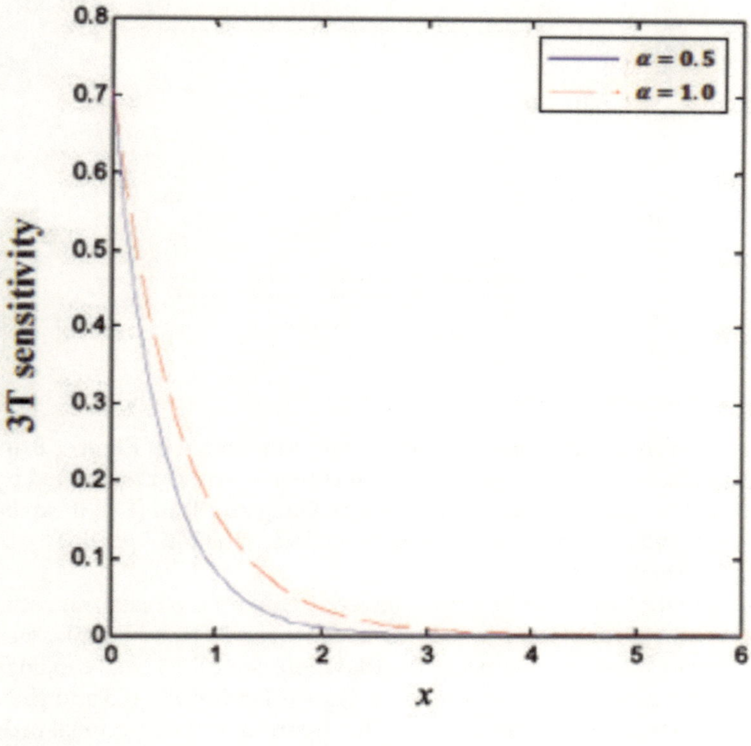

Figure 2.
Variation of the total 3T sensitivity along x-axis.

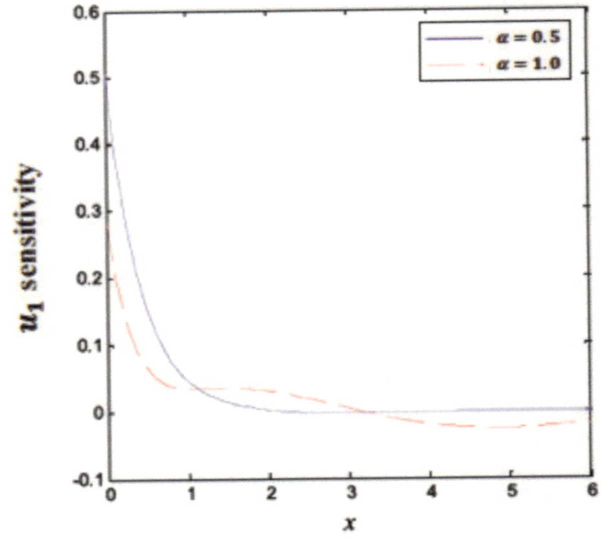

Figure 3.
Variation of the displacement u_1 *sensitivity along* x*-axis.*

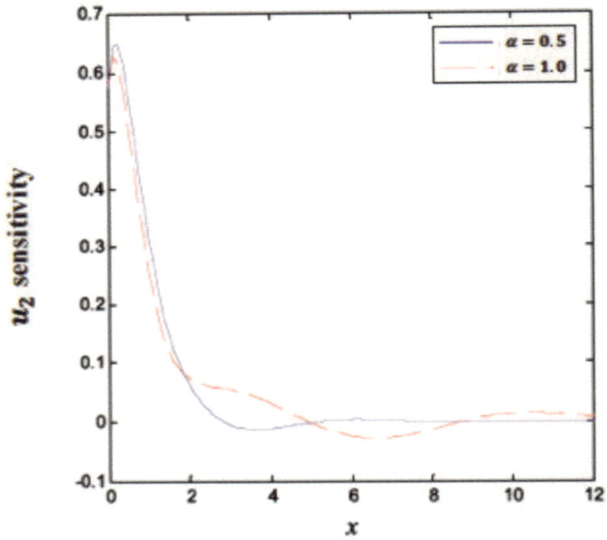

Figure 4.
Variation of the displacement u_2 *sensitivity along* x*-axis.*

the displacement u_1 and thermal stress σ_{11} results are plotted in **Figures 8** and **9**. The validity and accuracy of our proposed BEM technique were demonstrated by comparing our BEM results with the FEM results of Xiong and Tian [81], it can be noticed that the BEM results are found to agree very well with the FEM results.

Example 1. Short cantilever beam.

The mean compliance has been minimized, to obtain the maximum stiffness, when the structure is subjected to moving heat source. In this example, we consider a short cantilever beam shown in **Figure 10**, where the BESO final topology of considered short cantilever beam shown in **Figure 11a** for $\alpha = 0.5$ and shown in **Figure 11b** for $\alpha = 1.0$. It is noticed from this figure that the fractional order parameter has a significant effect on the final topology of the multi-material ISMFGA structure.

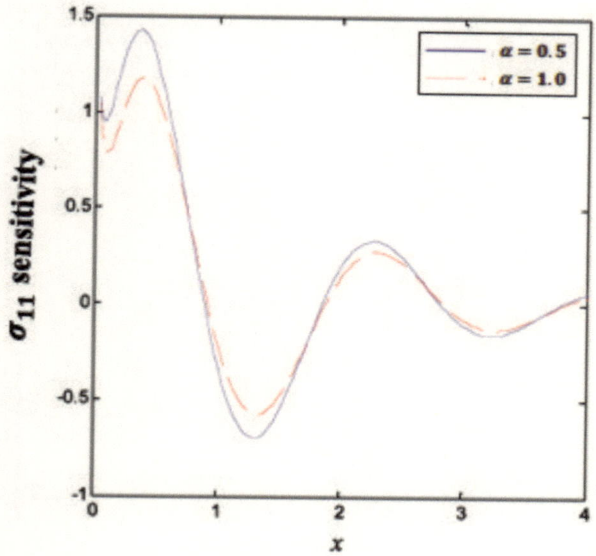

Figure 5.
Variation of the thermal stress σ_{11} sensitivity along x*-axis.*

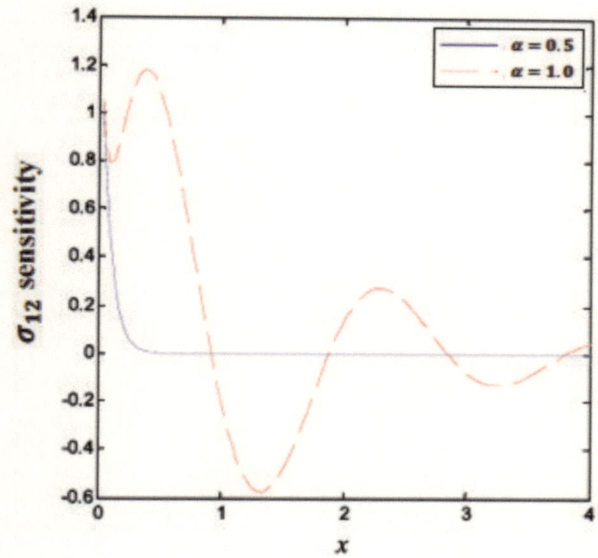

Figure 6.
Variation of the thermal stress σ_{12} sensitivity along x*-axis.*

Example 2. MBB beam.

It is known that extraordinary thermo-mechanical properties can be accomplished by combining more than two materials phases with conventional materials [75]. For this reason, it is essential that the topology optimization strategy permits more than two materials phases within the design domain. In this example, we consider a MBB beam shown in **Figure 12**, where the BESO final topology of MBB beam has been shown in **Figure 13a** for $\alpha = 0.5$ and shown in **Figure 13b** for $\alpha = 1.0$ to show the effect of fractional order parameter on the final topology of the multi-material ISMFGA structure.

Example 3. Roller-supported beam.

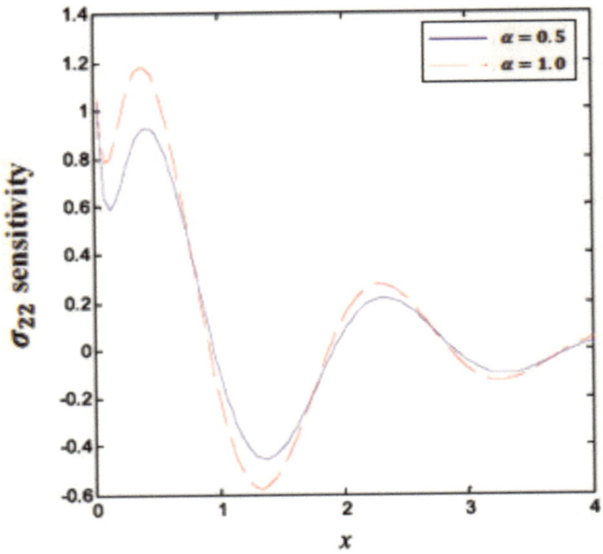

Figure 7.
Variation of the thermal stress σ_{22} *sensitivity along* x*-axis.*

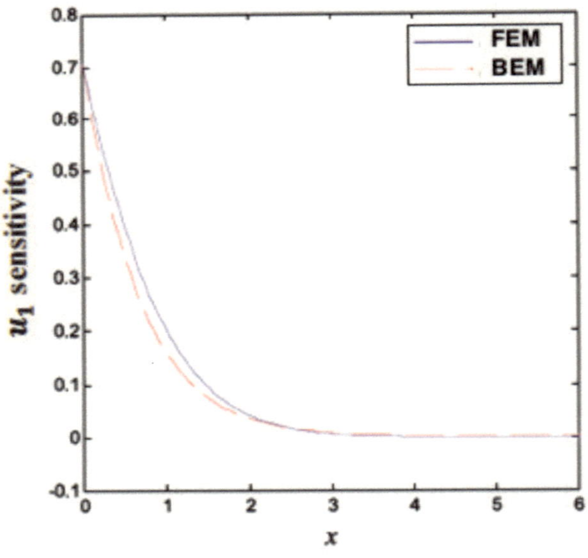

Figure 8.
Variation of the displacement u_1 *sensitivity along* x*-axis.*

In this example, we consider a roller-supported beam shown in **Figure 14**, where the BESO final topology of a roller-supported beam shown in **Figure 15a** for $\alpha = 0.5$ and shown in **Figure 15b** for $\alpha = 1.0$.

Example 4. Cantilever beam (validation example).

In order to demonstrate the validity of our implemented BESO topology optimization technique, we consider isotropic case of a cantilever beam shown in **Figure 16** as a special case of our anisotropic study to interpolate the elasticity matrix and the stress-temperature coefficients using the design variables X^M, then we compare our

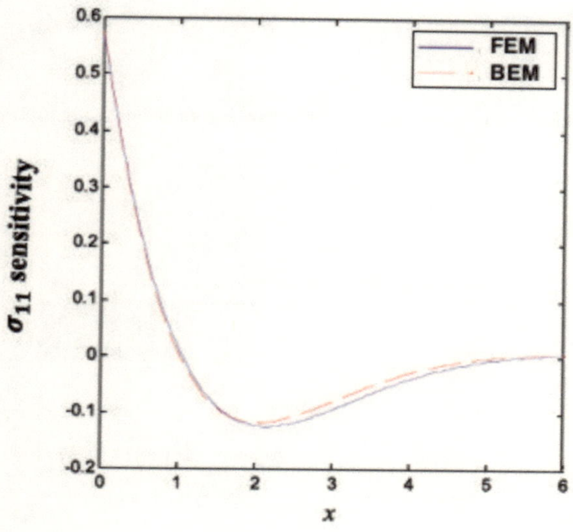

Figure 9.
Variation of the thermal stress σ_{11} waves along x-axis.

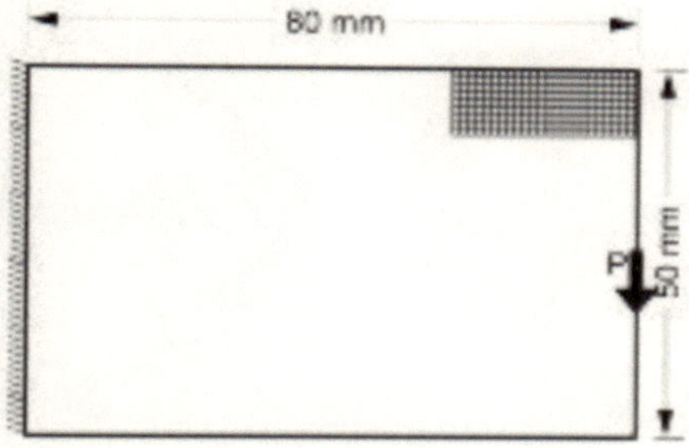

Figure 10.
Design domain of a short cantilever beam.

Figure 11.
The final topology of a short cantilever beam: (a) $\alpha = 0.5$ and (b) $\alpha = 1.0$.

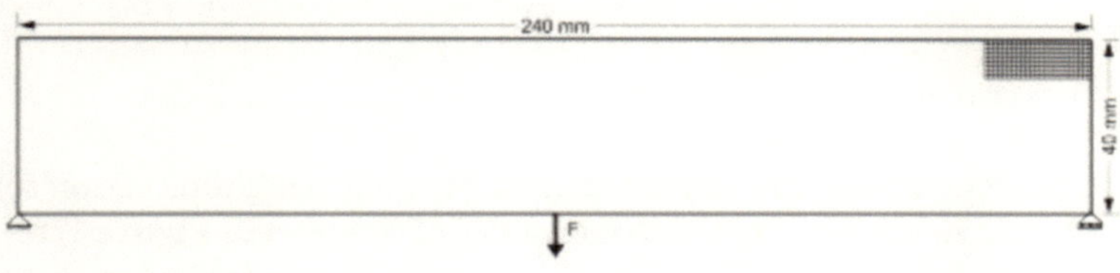

Figure 12.
Design domain of a MBB beam.

Figure 13.
The final topology of MBB beam: (a) α = 0.5 and (b) α = 1.0.

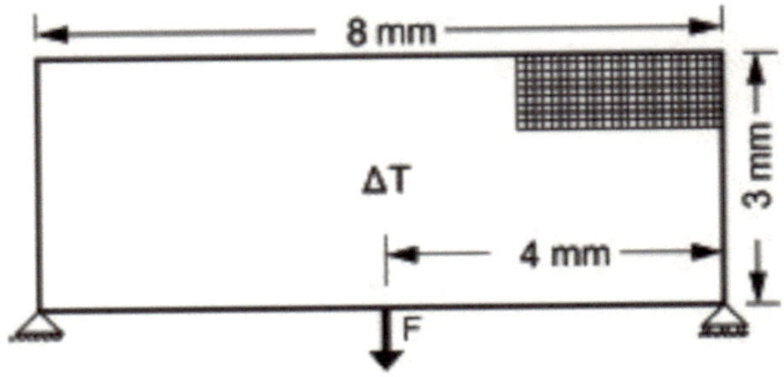

Figure 14.
Design domain of a roller-supported beam.

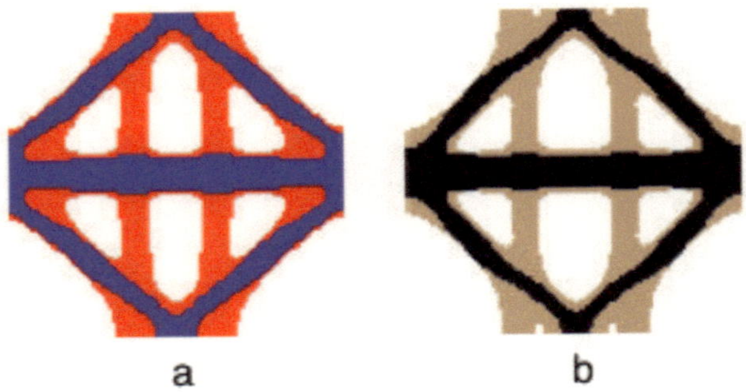

Figure 15.
The final topology of a multi-material roller-supported beam: (a) α = 0.5 and (b) α = 1.0.

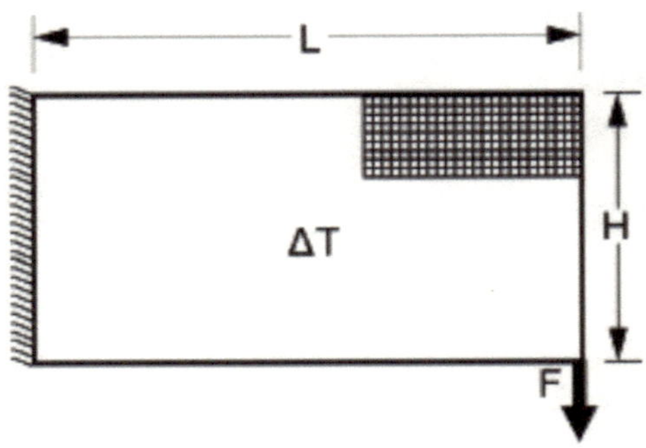

Figure 16.
Design domain of a cantilever beam.

BESO final topology shown in **Figure 17a** with the material interpolation scheme of the solid isotropic material with penalization (SIMP) shown in **Figure 17b**.

The BESO topology optimization problem implemented in Examples 1 and 4, to find the distribution of the M material phases, with the volume constraint can be stated as

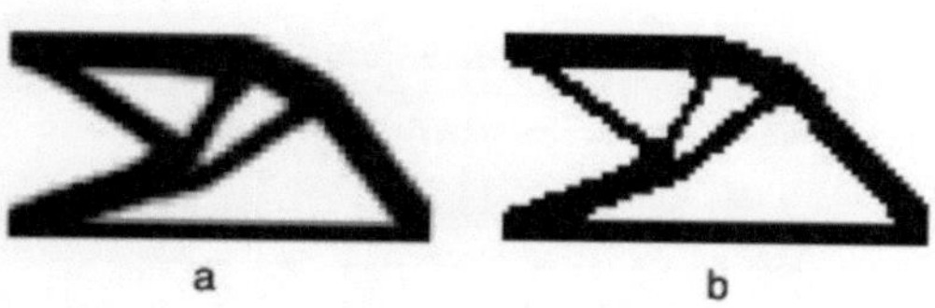

Figure 17.
The final topology of a cantilever beam: (a) MMA and (b) BESO.

Find X^M

That minimize $C^M = \frac{1}{2}(P^M)^T u^M = \frac{1}{2}\left(f^{M,ter} + f^{M,mec}\right)^T u^M$

Subject to $V^{M,^*} - \sum_{i=1}^{N} V_i^M X_i^M = 0$

$$K^M u^M = P^M$$

$$X_i^M = x_{min} \mathrm{V1}$$

where X^M is the design variable; C^M is the mean compliance; P is the total load on the structure, which is the sum of mechanical and thermal loads; u^M is the displacement vector; $V^{M,^*}$ is the volume of the solid material; N is the total number of elements; K^M is the global stiffness matrix; x_{min} is a small value (e.g., 0.0001), which it guarantee that none of the elements will be removed completely from design domain; $f^{M,mec}$ is the mechanical load vector; and $f^{M,ter}$ is the thermal load vector. Also, the BESO parameters considered in Examples 1–4 can be seen in **Tables 1–4**, respectively.

The BESO topology optimization problem implemented in Examples 2 and 3, to find the distribution of the two materials in the design domain, which minimize the compliance of the structure, subject to a volume constraint in both phases can be stated as.

Find X^M

That minimize $C^M = \frac{1}{2}(P^M)^T u^M = \frac{1}{2}\left(f^{M,ter} + f^{M,mec}\right)^T u^M$

Subject to $V_j^{M,^*} - \sum_{i=1}^{N} V_i^M X_{ij}^M - \sum_{i=1}^{j-1} V_i^{M,^*} = 0; j = 1, 2$

$$K^M u^M = P^M$$

$$X_i^M = x_{min} \mathrm{V1}; j = 1, 2$$

Variable name	Variable description	Variable value
V_f^M	Final volume fraction	0.5
ER^M	Evolutionary ratio	1%
AR_{max}^M	Volume addition ratio	5%
r_{min}^M	Filter ratio	3 mm
τ	Convergence tolerance	0.1%
N	Convergence parameter	5

Table 1.
BESO parameters for minimization of a short cantilever beam.

Variable name	Variable description	Variable value
V_{f1}^{M}	Final volume fraction of the material 1 for both interpolations	0.10
V_{f2}^{M}	Final volume fraction of the material 2 for both interpolations	0.20
ER^{M}	Evolutionary ratio for interpolation 1	2%
ER^{M}	Evolutionary ratio for interpolation 2	3%
AR_{max}^{M}	Volume addition ratio for interpolation 1	3%
AR_{max}^{M}	Volume addition ratio for interpolation 2	2%
r_{min}^{M}	Filter ratio for interpolation 1	4 mm
r_{min}^{M}	Filter ratio for interpolation 2	3 mm
τ	Convergence tolerance for both interpolations	0.01%
N	Convergence parameter for both interpolations	5

Table 2.
Multi-material BESO parameters for minimization of a MBB beam.

Variable name	Variable description	Variable value
V_{f1}^{M}	Final volume fraction of the material 1 for both interpolations	0.25
V_{f1}^{M}	Final volume fraction of the material 2 for both interpolations	0.25
ER^{M}	Evolutionary ratio for interpolation 1	3%
ER^{M}	Evolutionary ratio for interpolation 2	3%
AR_{max}^{M}	Volume addition ratio for interpolation 1	1%
AR_{max}^{M}	Volume addition ratio for interpolation 2	1%
r_{min}^{M}	Filter ratio for interpolation 1	4 mm
r_{min}^{M}	Filter ratio for interpolation 2	4 mm
τ	Convergence tolerance for both interpolations	0.5 %
N	Convergence parameter for both interpolations	5

Table 3.
Multi-material BESO parameters for minimization of a roller-supported beam.

Variable name	Variable description	Variable value
V_{f}^{M}	Final volume fraction	0.4
ER^{M}	Evolutionary ratio	1.2%
AR_{max}^{M}	Volume addition ratio	3%
r_{min}^{M}	Filter ratio	0.19 mm
τ	Convergence tolerance	0.1%
N	Convergence parameter	5

Table 4.
BESO parameters for minimization of a cantilever beam.

where $V_j^{M^*}$ is the volume of jth material phase and i and j denote the element ith which is made of jth material.

6. Conclusion

The main purpose of this chapter is to describe a new boundary element formulation for modeling and optimization of 3T time fractional order nonlinear generalized thermoelastic multi-material ISMFGA structures subjected to moving heat source, where we used the three-temperature nonlinear radiative heat conduction equations combined with electron, ion, and phonon temperatures.

Numerical results show the influence of fractional order parameter on the sensitivities of the study's fields. The validity of the present method is examined and demonstrated by comparing the obtained outcomes with those known in the literature. Because there are no available data to confirm the validity and accuracy of our proposed technique, we replace the three-temperature radiative heat conduction with one-temperature heat conduction as a special case from our current general study of three-temperature nonlinear generalized thermoelasticity. In the considered special case of 3T time fractional order nonlinear generalized thermoelastic multi-material ISMFGA structures, the BEM results have been compared graphically with the FEM results; it can be noticed that the BEM results are in excellent agreement with the FEM results. These results thus demonstrate the validity and accuracy of our proposed technique. Numerical examples are solved using the multi-material topology optimization algorithm based on the bi-evolutionary structural optimization method (BESO). Numerical results of these examples show that the fractional order parameter affects the final result of optimization. The implemented optimization algorithm has proven to be an appropriate computational tool for material design.

Nowadays, the knowledge of 3T fractional order optimization of multi-material ISMFGA structures, can be utilized by mechanical engineers for designing heat exchangers, semiconductor nano materials, thermoelastic actuators, shape memory actuators, bimetallic valves and boiler tubes. As well as for chemists to observe the chemical processes such as bond breaking and bond forming.

Author details

Mohamed Abdelsabour Fahmy
Faculty of Computers and Informatics, Suez Canal University, Ismailia, Egypt

*Address all correspondence to: mohamed_fahmy@ci.suez.edu.eg

References

[1] Cattaneo C. Sur une forme de i'equation de la chaleur elinant le paradox d'une propagation instantanc. Comptes Rendus de l'Académie des Sciences. 1958;**247**:431-433

[2] Oldham KB, Spanier J. The Fractional Calculus: Theory and Applications of Differentiation and Integration to Arbitrary Order. Mineola: Dover Publication; 2006

[3] Podlubny I. Fractional Differential Equations. San Diego, California, USA: Academic Press; 1999

[4] Ezzat MA, El Karamany AS, Fayik MA. Fractional order theory in thermoelastic solid with three-phase lag heat transfer. Archive of Applied Mecha nics. 2012;**82**:557-572

[5] Kilbas AA, Srivastava HM, Trujillo JJ. Theory and Applications of Fractional Differential Equations, vol. 204 of North-Holland Mathematics Studies. Amsterdam, The Netherlands: Elsevier Science; 2006

[6] Sabatier J, Agrawal OP, Machado JAT. Advances in Fractional Calculus: Theoretical Developments and Applications in Physics and Engineering. Dordrecht, The Netherlands: Springer; 2007

[7] Ezzat MA, El-Bary AA. Application of fractional order theory of magneto-thermoelasticity to an infinite perfect conducting body with a cylindrical cavity. Microsystem Technologies. 2017; **23**:2447-2458

[8] Soukkou A, Belhour MC, Leulmi S. Review, design, optimization and stability analysis of fractional-order PID controller. International Journal of Intelligent Systems Technologies and Applications. 2016;**8**:73-96

[9] El-Naggar AM, Abd-Alla AM, Fahmy MA. The propagation of thermal stresses in an infinite elastic slab. Applied Mathematics and Computation. 2003;**12**:220-226

[10] Fahmy MA. A new boundary element strategy for modeling and simulation of three temperatures nonlinear generalized micropolar-magneto-thermoelastic wave propagation problems in FGA structures. Engineering Analysis with Boundary Elements. 2019;**108**:192-200

[11] Fahmy MA. A new computerized boundary element model for three-temperature nonlinear generalized thermoelastic stresses in anisotropic circular cylindrical plate structures. In: Awrejcewicz J, Grzelczyk D, editors. Dynamical Systems Theory. London, UK: IntechOpen; 2019. pp. 1-17

[12] Fahmy MA. Boundary element model for nonlinear fractional-order heat transfer in magneto-thermoelastic FGA structures involving three temperatures. In: Ebrahimi F, editor. Mechanics of Functionally Graded Materials and Structures. London, UK: IntechOpen; 2019. pp. 1-22

[13] Fahmy MA. Boundary element mathematical modelling and boundary element numerical techniques for optimization of micropolar thermoviscoelastic problems in solid deformable bodies. In: Sivasankaran S, Nayak PK, Günay E, editors. Mechanics of Solid Deformable Bodies. London, UK: IntechOpen; 2020. pp. 1-21

[14] Fahmy MA. Boundary element modeling and optimization based on fractional-order derivative for nonlinear generalized photo-thermoelastic stress wave propagation in three-temperature anisotropic semiconductor structures. In: Sadollah A, Sinha TS, editors. Recent Trends in Computational Intelligence. London, UK: IntechOpen; 2020. pp. 1-16

[15] Abd-Alla AM, El-Naggar AM, Fahmy MA. Magneto-thermoelastic problem in non-homogeneous isotropic cylinder. Heat and Mass Transfer. 2003; **39**:625-629

[16] Hu Q, Zhao L. Domain decomposition preconditioners for the system generated by discontinuous Galerkin discretization of 2D-3T heat conduction equations. Communications in Computational Physics. 2017;**22**: 1069-1100

[17] Sharma N, Mahapatra TR, Panda SK. Thermoacoustic behavior of laminated composite curved panels using higher-order finite-boundary element model. International Journal of Applied Mechanics. 2018;**10**:1850017

[18] Fahmy MA. A time-stepping DRBEM for magneto-thermo-viscoelastic interactions in a rotating nonhomogeneous anisotropic solid. International Journal of Applied Mechanics. 2011;**3**:1-24

[19] Fahmy MA. A time-stepping DRBEM for the transient magneto-thermo-visco-elastic stresses in a rotating non-homogeneous anisotropic solid. Engineering Analysis with Boundary Elements. 2012;**36**:335-345

[20] Fahmy MA. Numerical modeling of transient magneto-thermo-viscoelastic waves in a rotating nonhomogeneous anisotropic solid under initial stress. International Journal of Modeling, Simulation and Scientific Computing. 2012;**3**:1250002

[21] Fahmy MA. Transient magneto-thermo-viscoelastic stresses in a rotating nonhomogeneous anisotropic solid with and without a moving heat source. Journal of Engineering Physics and Thermophysics. 2012;**85**:950-958

[22] Fahmy MA. Transient magneto-thermo-elastic stresses in an anisotropic viscoelastic solid with and without moving heat source. Numerical Heat Transfer, Part A: Applications. 2012;**61**: 547-564

[23] Fahmy MA. Transient magneto-thermoviscoelastic plane waves in a non-homogeneous anisotropic thick strip subjected to a moving heat source. Applied Mathematical Modelling. 2012; **36**:4565-4578

[24] Fahmy MA. The effect of rotation and inhomogeneity on the transient magneto-thermoviscoelastic stresses in an anisotropic solid. ASME Journal of Applied Mechanics. 2012;**79**:1015

[25] Fahmy MA. Computerized Boundary Element Solutions for Thermoelastic Problems: Applications to Functionally Graded Anisotropic Structures. Saarbrücken: LAP Lambert Academic Publishing; 2017

[26] Fahmy MA. Boundary Element Computation of Shape Sensitivity and Optimization: Applications to Functionally Graded Anisotropic Structures. Saarbrücken: LAP Lambert Academic Publishing; 2017

[27] Fahmy MA. A time-stepping DRBEM for 3D anisotropic functionally graded piezoelectric structures under the influence of gravitational waves. In: Proceedings of the 1st GeoMEast International Congress and Exhibition (GeoMEast 2017); 15–19 July 2017; Sharm El Sheikh, Egypt. Facing the Challenges in Structural Engineering, Sustainable Civil Infrastructures. 2017. pp. 350-365

[28] Fahmy MA. 3D DRBEM modeling for rotating initially stressed anisotropic functionally graded piezoelectric plates. In: Proceedings of the 7th European Congress on Computational Methods in Applied Sciences and Engineering (ECCOMAS 2016); 5–10 June 2016; Crete Island, Greece. 2016. pp. 7640-7658

[29] Fahmy MA. Boundary element solution of 2D coupled problem in

anisotropic piezoelectric FGM plates. In: Proceedings of the 6th International Conference on Computational Methods for Coupled Problems in Science and Engineering (Coupled Problems 2015); 18–20 May 2015; Venice, Italy. 2015. pp. 382-391

[30] Fahmy MA. The DRBEM solution of the generalized magneto-thermo-viscoelastic problems in 3D anisotropic functionally graded solids. In: Proceedings of the 5th International Conference on Coupled Problems in Science and Engineering (Coupled Problems 2013); 17–19 June 2013; Ibiza, Spain. 2013. pp. 862-872

[31] Fahmy MA. A computerized boundary element model for simulation and optimization of fractional-order three temperatures nonlinear generalized piezothermoelastic problems based on genetic algorithm. In: AIP Conference Proceedings 2138 of Innovation and Analytics Conference and Exihibiton (IACE 2019); 25-28 March 2019; Sintok, Malaysia. 2019. p. 030015

[32] Soliman AH, Fahmy MA. Range of applying the boundary condition at fluid/porous Interface and evaluation of beavers and Joseph's slip coefficient using finite element method. Computation. 2020;**8**:14

[33] Eskandari AH, Baghani M, Sohrabpour S. A time-dependent finite element formulation for thick shape memory polymer beams considering shear effects. International Journal of Applied Mechanics. 2019;**10**:1850043

[34] Huang R, Zheng SJ, Liu ZS, Ng TY. Recent advances of the constitutive models of smart materials—Hydrogels and shape memory polymers. International Journal of Applied Mechanics. 2020;**12**:2050014

[35] Othman MIA, Khan A, Jahangir R, Jahangir A. Analysis on plane waves through magneto-thermoelastic microstretch rotating medium with temperature dependent elastic properties. Applied Mathematical Modelling. 2019;**65**:535-548

[36] El-Naggar AM, Abd-Alla AM, Fahmy MA, Ahmed SM. Thermal stresses in a rotating non-homogeneous orthotropic hollow cylinder. Heat and Mass Transfer. 2002;**39**:41-46

[37] Fahmy MAA. New computerized boundary element algorithm for cancer modeling of cardiac anisotropy on the ECG simulation. Asian Journal of Research in Computer Science. 2018;**2**: 1-10

[38] Brebbia CA, Telles JCF, Wrobel L. Boundary Element Techniques in Engineering. New York: Springer-Verlag; 1984

[39] Wrobel LC, Brebbia CA. The dual reciprocity boundary element formulation for nonlinear diffusion problems. Computer Methods in Applied Mechanics and Engineering. 1987;**65**:147-164

[40] Partridge PW, Wrobel LC. The dual reciprocity boundary element method f or spontaneous ignition. International Journal for Numerical Methods in Engineering. 1990;**30**:953-963

[41] Partridge PW, Brebbia CA. Computer implementation of the BEM dual reciprocity method for the solution of general field equations. Communications in Applied Numerical Methods. 1990;**6**:83-92

[42] Partridge PW, Brebbia CA, Wrobel LC. The Dual Reciprocity Boundary Element Method. Southampton: Computational Mechanics Publications; 1992

[43] Gaul L, Kögl M, Wagner M. Boundary Element Methods for

Engineers and Scientists. Berlin: Springer-Verlag; 2003

[44] Fahmy MA. Implicit-explicit time integration DRBEM for generalized magneto-thermoelasticity problems of rotating anisotropic viscoelastic functionally graded solids. Engineering Analysis with Boundary Elements. 2013; **37**:107-115

[45] Fahmy MA. Generalized magneto-thermo-viscoelastic problems of rotating functionally graded anisotropic plates by the dual reciprocity boundary element method. Journal of Thermal Stresses. 2013;**36**:1-20

[46] Fahmy MA. A three-dimensional generalized magneto-thermo-viscoelastic problem of a rotating functionally graded anisotropic solids with and without energy dissipation. Numerical Heat Transfer, Part A: Applications. 2013;**63**:713-733

[47] Fahmy MA. A 2-D DRBEM for generalized magneto-thermo-viscoelastic transient response of rotating functionally graded anisotropic thick strip. International Journal of Engineering and Technology Innovation. 2013;**3**:70-85

[48] Fahmy MA, Salem AM, Metwally MS, Rashid MM. Computer implementation of the DRBEM for studying the generalized Thermoelastic responses of functionally graded anisotropic rotating plates with one relaxation time. International Journal of Applied Science and Technology. 2013; **3**:130-140

[49] Fahmy MA, Salem AM, Metwally MS, Rashid MM. Computer implementation of the DRBEM for studying the classical uncoupled theory of thermoelasticity of functionally graded anisotropic rotating plates. International Journal of Engineering Research and Applications. 2013;**3**: 1146-1154

[50] Fahmy MA. A computerized DRBEM model for generalized magneto-thermo-visco-elastic stress waves in functionally graded anisotropic thin film/substrate structures. Latin American Journal of Solids and Structures. 2014;**11**:386-409

[51] Fahmy MA, Salem AM, Metwally MS, Rashid MM. Computer implementation of the DRBEM for studying the classical coupled thermoelastic responses of functionally graded anisotropic plates. Physical Science International Journal. 2014;**4**: 674-685

[52] Fahmy MA, Salem AM, Metwally MS, Rashid MM. Computer implementation of the DRBEM for studying the generalized thermo elastic responses of functionally graded anisotropic rotating plates with two relaxation times. British Journal of Mathematics & Computer Science. 2014;**4**:1010-1026

[53] Fahmy MA. A 2D time domain DRBEM computer model for magneto-thermoelastic coupled wave propagation problems. International Journal of Engineering and Technology Innovation. 2014;**4**:138-151

[54] Fahmy MA, Al-Harbi SM, Al-Harbi BH. Implicit time-stepping DRBEM for design sensitivity analysis of magneto-thermo-elastic FGA structure under initial stress. American Journal of Mathematical and Computational Sciences. 2017;**2**:55-62

[55] Fahmy MA. The effect of anisotropy on the structure optimization using golden-section search algorithm based on BEM. Journal of Advances in Mathematics and Computer Science. 2017;**25**:1-18

[56] Fahmy MA. DRBEM sensitivity analysis and shape optimization of rotating magneto-thermo-viscoelastic FGA structures using golden-section

search algorithm based on uniform bicubic B-splines. Journal of Advances in Mathematics and Computer Science. 2017;**25**:1-20

[57] Fahmy MA. A predictor-corrector time-stepping DRBEM for shape design sensitivity and optimization of multilayer FGA structures. Transylvanian Review. 2017;**XXV**:5369-5382

[58] Fahmy MA. Shape design sensitivity and optimization for two-temperature generalized magneto-thermoelastic problems using time-domain DRBEM. Journal of Thermal Stresses. 2018;**41**: 119-138

[59] Fahmy MA. Boundary element algorithm for modeling and simulation of dual-phase lag bioheat transfer and biomechanics of anisotropic soft tissues. International Journal of Applied Mechanics. 2018;**10**:1850108

[60] Fahmy MA. Shape design sensitivity and optimization of anisotropic functionally graded smart structures using bicubic B-splines DRBEM. Eng ineering Analysis with Boundary Eleme nts. 2018;**87**:27-35

[61] Fahmy MA. Modeling and optimization of anisotropic viscoelastic porous structures using CQBEM and moving asymptotes algorithm. Arabian Journal for Science and Engineering. 2019;**44**:1671-1684

[62] Fahmy MA. Boundary element modeling and simulation of biothermomechanical behavior in anisotropic laser-induced tissue hyperthermia. Engineering Analysis with Boundary Elements. 2019;**101**: 156-164

[63] Fahmy MA, Al-Harbi SM, Al-Harbi BH, Sibih AM. A computerized boundary element algorithm for modeling and optimization of complex magneto-thermoelastic problems in MFGA structures. Journal of Engineering Research and Reports. 2019;**3**:1-13

[64] Fahmy MA. A new LRBFCM-GBEM modeling algorithm for general solution of time fractional order dual phase lag bioheat transfer problems in functionally graded tissues. Numerical Heat Transfer, Part A: Applications. 2019;**75**:616-626

[65] Fahmy MA. Design optimization for a simulation of rotating anisotropic viscoelastic porous structures using time-domain OQBEM. Mathematics and Computers in Simulation. 2019;**66**: 193-205

[66] Fahmy MA. A new convolution variational boundary element technique for design sensitivity analysis and topology optimization of anisotropic thermo-poroelastic structures. Arab Journal of Basic and Applied Sciences. 2020;**27**:1-12

[67] Fahmy MA. Thermoelastic stresses in a rotating non- homogeneous anisotropic body. Numerical Heat Transfer, Part A: Applications. 2008;**53**: 1001-1011

[68] Abd-Alla AM, Fahmy MA, El-Shahat TM. Magneto-thermo-elastic problem of a rotating non-homogeneous anisotropic solid cylinder. Archive of Applied Mechanics. 2008;**78**:135-148

[69] Fahmy MA, El-Shahat TM. The effect of initial stress and inhomogeneity on the thermoelastic stresses in a rotating anisotropic solid. Archive of Applied Mechanics. 2008, 2008;**78**:431-442

[70] Huang X, Xie Y. Convergent and mesh-independent solutions for the bi-directional evolutionary structural optimization method. Finite Elements in Analysis and Design. 2007;**43**(14):1039-1049

[71] Huang X, Xie Y. Evolutionary Topology Optimization of Continuum Structures. U.S.A: John Wiley & Sons Ltd.; 2010

[72] Huang X, Xie YM. Bi-directional evolutionary topology optimization of continuum structures with one or multiple materials. Computational Mechanics. 2008;**43**(3):393

[73] Huang X, Zhou S, Xie Y, Li Q. Topology optimization of microstructures of cellular materials and composites for macrostructures. Computational Materials Science. 2013; **67**:397-407

[74] Sigmund O. Design of multiphysics actuators using topology optimization—Part i: One-material structures. Computer Methods in Applied Mechanics and Engineering. 2001; **190**(49):6577-6604

[75] Sigmund O, Torquato S. Composites with extremal thermal expansion coefficients. Applied Physics Letters. 1996;**69**(21):3203-3205

[76] Sigmund O, Torquato S. Design of materials with extreme thermal expansion using a three-phase topology optimization method. Journal of the Mechanics and Physics of Solids. 1997; **45**(6):1037-1067

[77] Sigmund O, Torquato S. Design of smart composite materials using topology optimization. Smart Materials and Structures. 1999;**8**:365-379

[78] Yang XY, Xei YM, Steven GP, Querin OM. Bidirectional evolutionary method for stiffness optimization. AIAA Journal. 1999;**37**(11):1483-1488

[79] Wang Y, Luo Z, Zhang N, Wu T. Topological design for mechanical metamaterials using a multiphase level set method. Structural and Multidisciplinary Optimization. 2016b; **54**:937-954

[80] Xu B, Huang X, Zhou S, XIE Y. Concurrent topological design of composite thermoelastic macrostructure and microstructure with multi-phase material for maximum stiffness. Composite Structures. 2016;**150**:84-102

[81] Xiong QL, Tian XG. Generalized magneto-thermo-microstretch response during thermal shock. Latin American Journal of Solids and Structures. 2015; **12**:2562-2580

Chapter 5

Using Algebraic Fractals in Steganography

Oleg Sheluhin and Dzhennet Magomedova

Abstract

Steganography is a technology for hiding watermarks inside media files, which is relevant in the field of copyright protection, secret communication, etc. The effectiveness of modern methods of digital image processing allows determining the presence of embedded watermarks in a stegoimage using the original image and its statistical characteristics, as well as a priori information about the method and algorithm of embedding. In contrast to the known approaches, it is proposed to use algebraic fractals for steganographic embedding of watermarks in color images. It is proposed to use algebraic fractals in the form of medium cover image acting as a secret key, which allows the embedding to be more resistant to computer attacks, including JPEG compression. The main advantage of such use of fractals is an increase in the level of secrecy in which the attacker must know the parameters of the fractal image. Without knowledge of these parameters, it will not be enough to have the original stegoimage and a priori information about the embedding method to extract secret data. This chapter analyzes the methods and provides examples of generating algebraic fractals in the form of the Julia set using the escape time algorithm.

Keywords: steganography, watermarks, color images, copyright protection, Julia set, escape time algorithm

1. Introduction

The digital form of audio, image, and video has become a commercial standard in the past decade. Digitized multimedia files can be easily created, copied, processed, saved, and distributed using commercial and free software. Unfortunately, the digitization of multimedia files leads to the fact that these files are subject to digital piracy: illegal copying, use, and distribution of copyrighted digital data. In order to combat digital piracy, various copyright protection mechanisms have been developed for many years. However, most of these mechanisms were found to be erroneous and unsafe.

One of the tools to protect multimedia data from copyright infringement is digital watermarks. Into digital watermarks, an imperceptible "mark" signal is embedded in the original image. This label uniquely identifies the owner. After embedding the watermark in the original image (container) there should be no noticeable distortion. Embedded watermarks should not be removed by an incomplete person and must be resistant to intentional and unintentional attacks.

One of the ways to protect copyright, which currently exists, is steganography, the art of hiding information in electronic media. Methods of steganography vary in their approach to hiding information. As a rule, the hidden data in electronic means will change some of their properties, which can lead to degradation or unusual characteristics.

The paper deals with the introduction of watermark into a still color image of JPEG format using steganography methods. A feature of the proposed method is the use of a fractal image as a secret key, in which a two-dimensional fractal of algebraic type in the form of a Julia set is used. As a result, an evil intentional person will not be able to generate an identical fractal image without the exact value of a certain complex number, which is agreed in advance between the sender and the recipient, which makes the proposed method resistant to attacks.

Algorithms for creating a fractal image are considered, into which a digital watermark is then embedded.

To embed the fractal image in the container, the algorithm that performs a three-level expansion using the Haar transform was chosen and implemented in the MATLAB. In order to optimize the quality of extraction and the amount of embedded information, an algorithm for calculating thresholds has been proposed, with the help of which the necessary number of wavelet decomposition coefficients suitable for embedding has been selected.

To assess the possibility of detecting the fact of embedding the watermark in the stegoimage, the method of estimating the fractal dimension (FR) can be used before and after the mark has been embedded. With numerous examples of parameters of algebraic fractals and the hidden algorithms used, it is shown that the introduction of a watermark slightly changes the fractal dimension of the stegoimage, which corresponds to the high stability of the method to possible steganographic attacks.

To assess the quality of the results obtained, the quality was estimated based on the following metrics: mean square error (MSE) and signal-to-noise ratio (SNR), expressed in decibels. The results of the original container and the container in which the watermark with the key in the form of a fractal are embedded, as well as the original and extracted watermark, are given.

Analysis of the results of the quality assessment of the original container and the container with the embedded watermark and the key in the form of a fractal, as well as the original and extracted watermark showed that the proposed algorithm provides a high quality of hiding confidential information.

2. Algorithm development and software implementation of the generation of a fractal image of Julia set

The fractal image of a Julia sets will form using an algorithm of time escape (Escape time algorithm) [1, 2]. The algorithm is based on the use of complex maps when one complex number $z_n = x_n + iy_n$ is matched by another complex number $z_{n+1} = x_{n+1} + iy_{n+1}$ according to the iterative rule $z_{n+1} = f(z_n)$, where f(z) is some nonlinear function of the argument z, and n is the iteration number.

The algorithm uses a quadratic complex polynomial: $f(z) = z^2 + c$ where $c = x + iy$ is the starting point on the complex plane on the basis of which the Julia set is constructed.

If point c belongs to the Mandelbrot set, the Julia set constructed on it is connected [3]. If c does not belong to the Mandelbrot set, the constructed sets are scattered into an infinite number of isolated points (Fatu dust). If the point c lies near the boundary of the Mandelbrot set, such sets form fractal figures figuratively called the "seahorse valley" in the vicinity of zo = 0, which have the property of self-similarity with chaotic dynamics.

Thus, the algorithm for constructing the Julia set takes the form of [4]:

1. Select point c to set polynomial $f(z) = z^2 + c$;

2. Calculate the radius R for a given polynomial $f(z) = z^2 + c$;

3. Select the maxstep parameter to indicate the maximum iteration. The higher it is, the higher the accuracy and the slower the algorithm;

4. Generate an array of colors from less bright to brighter. We will denote the color dependence of the removal of points from a variety of Julia sets;

5. For each point, we calculate whether it is part of the filled Julia set or not, as well as the iteration number at which the threshold has exceeded. If |z| > R then use the first color, then use the color palette on which the iteration number was exceeded a threshold.

The implementation of the algorithm begins with the construction of the rectangle L × L, within which the Julia set will be constructed. By changing its parameters, we can "bring closer" the concentrations of interest to a set with high accuracy. The resulting rectangle is divided into parts with coordinates $z_{i,j}$ depending on the number of pixels of the image M × M.

Each point $z_{i,j}$ enters an iterative loop $f(z) = z^2 + c$. The number of iterations depends on the maxstep parameter. If after a given number of iterations a point does not go to infinity, then such a point belongs to the Julia set.

Depending on which area of the Julia set will be generated (L × L), a certain number of points will go to infinity, without reaching the maximum iteration.

To optimize the algorithm and significantly reduce the generation time of the Julia set, the following approach was adopted: for a complex number, the radius $R = \frac{1}{2} + \frac{\sqrt{1+4|c|}}{2}$ is calculated, and at each iteration, the test $z_{i,j} > R$ occurs, for points satisfying this condition, iterations are no longer performed.

The points $z_{i,j}$ extending beyond this radius will go to infinity through a certain number of iterations; since their attractor is an infinitely distant point (points $z_{i,j} > R$ enter the pool of the attractor A (∞)). The basin of an attractor of a point z of a function f is a subset of points from a neighborhood of z (denoted as A (z)), that any trajectory starting at one of these points converges to a point z.

Such an approach can significantly reduce the number of calculations, especially with the general form of the Julia set (2 × 2). However, with an increase in the approximation to the level (0.005 × 0.005) the number of calculations will tend to the initial ones, as more and more points will belong to the Julia Set. The color in the implementation of the algorithm was chosen so that the absolute red color (with a brightness level of 255) has points (belonging to the Julia set) that do not go to infinity with the maximum number of iterations maxstep. The brightness of the "red component" of the rest of the check depends on the number of iterations that the point passed, going to infinity (points not belonging to the Julia set, but rather close to its boundary). For the green channel, an inverse relationship was used: the smaller the iterations, the greater the brightness of the green component up to the level of 255, if the point was originally outside the radius. The blue channel changes depending on the distance of the point from the origin and is minimal at z = 0.

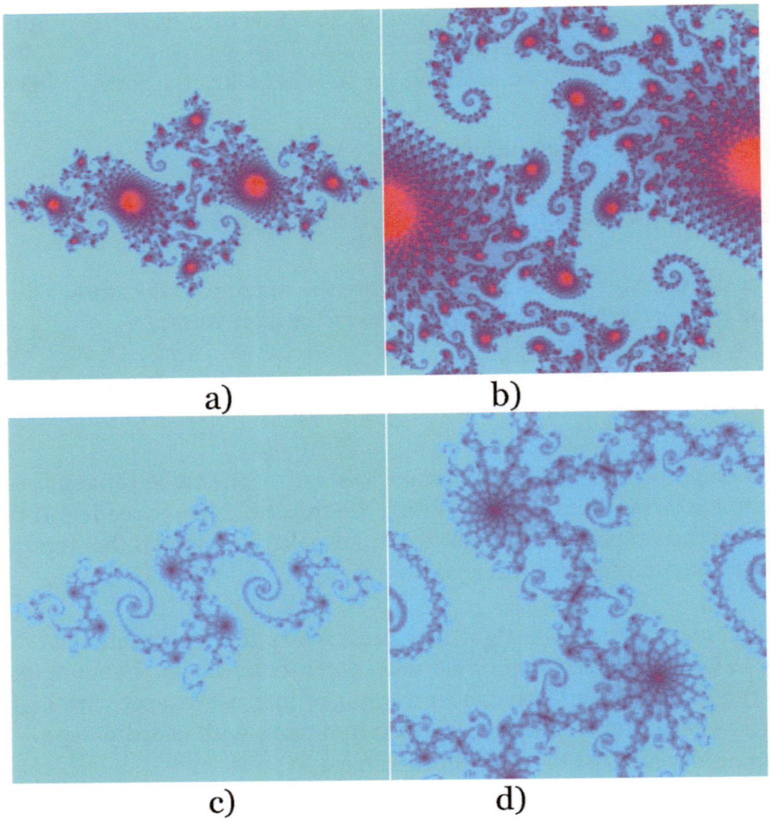

Figure 1.
Generated fractals with various parameters: (a) $c = -0.74543 + 0.11301i$, maxstep = 300, $l = 1.5$, $m = 1024$; (b) $c = -0.74543 + 0.11301i$, maxstep = 300, $l = 0.5$, $m = 1024$; (c) $c = -0.77780781 + 0.13164510i$, maxstep = 300, $l = 1.5$, $m = 1024$; and (d) $c = -0.77780781 + 0.13164510i$, maxstep = 300, $l = 0.5$, $m = 1024$.

The implementation of the above method in the MATLAB math package is shown in Listing 1:

```
clear;
cx=0;
cy=0;
l=0.005;
maxstep=300;
m=600;
max=0;
C=-0.74543 + 0.11301i;
r=0.5+sqrt(1+4*abs(C))/2;

for i=1:m
for j=1:m
z(i,j)=-l+j*2*l/(m-1)+(-l+i*2*l/(m-1))*1i;
for k=1:maxstep
w((i-1)*m+j)=maxstep;
if (abs(z(i,j))>r)
w((i-1)*m+j)=k;
break;
end
z(i,j)=power(z(i,j),2)+C;
```

```
  if w((i-1)*m+j)>max
  max=w((i-1)*m+j);
  end
  end
  Red(i,j)=uint8(255*w((i-1)*m+j)/max);
  Green(i,j)=uint8(255-255*w((i-1)*m+j)/max);
  if (abs(z(i,j))/r)>1
  Blue(i,j)=255;
  else
  Blue(i,j)=uint8(255*(abs(z(i,j)/r)));
  end
  end
 end
 img=Red;
 img(:,:,2)=Green;
 img(:,:,3)=Blue;
 imwrite(img,'fractal.bmp');
```

User-defined initial conditions:

maxstep - the maximum number of iterations, which determines the detailing of the Julia set;

l - the dimension of the rectangle (determines the scale of the approximation);

cx, cy- coordinates of the center of the rectangle (determine the coordinates of the approximation);

m is the size of the container (the number of pixels of the image mxm).

The results of the developed program module are presented in **Figure 1**.

3. Embedding a watermark in a stegoimage using a "fractal key"

The created fractal image based on Julia's set can be used as an intermediate cover image (key) for embedding a digital watermark into it by one of the well-known methods [5–7]. So, for example, when using the least significant bit method when embedding, a fractal is converted into a one-dimensional binary array and every eighth bit of the given array (that is, every least significant bit of the next fractal byte) is replaced with a watermark bit.

After the watermark is fully integrated into the created fractal image, the step of embedding the filled fractal key into the original image begins [8–10].

As an example, consider embedding a watermark in a container using a 2D wavelet algorithm. To embed a watermark (in our case, a fractal image containing a watermark) in a cover image using 2D fiberboard, the algorithm [11–13] was used as a basis, which uses a three-level wavelet decomposition using the Haar transform. The decomposition of the container in the area of coefficients of detail of the first level LH1, HH1, HL1 is as follows:

A horizontal Haar transform is applied (across all rows) to the source container (Eqs. (1) and (2)).

$$C1[i,jj] = \frac{Cont[i,j] + Cont[i,j+1]}{2} \tag{1}$$

$$C2[i,jj] = \frac{Cont[i,j] - Cont[i,j+1]}{2} \tag{2}$$

where $i = 1, \dots, M; j = 1, \dots, N; jj = j, \dots, \frac{N}{2}$.

In Eqs. (1) and (2), the following notation is used: Cont is an array of pixels of the original image; C1 is the half-sum matrix of pixels in the original image; C2 is the half-difference matrix of the pixels of the original image; M is the number of rows of the original image; and N is the number of columns of the original image.

The resulting matrices C1 and C2 will have the same dimension [i, jj], which means that the number of columns will be two times less relative to the Cont matrix of the original image. Then, using the values of the matrices C1 and C2, we calculate the coefficients of approximation (LL1) and details (HL1, LH1, HH1) for level 1 of the decomposition.

$$LL1[ii,jj] = C1[i,jj] + C1[i+1,jj] \tag{3}$$

$$HL1[ii,jj] = C1[i,jj] - C1[i+1,jj] \tag{4}$$

$$LH1[ii,jj] = C2[i,jj] + C2[i+1,jj] \tag{5}$$

$$HH1[ii,jj] = C2[i,jj] - C2[i+1,jj] \tag{6}$$

where $i = 1, \dots, M; j = 1, \dots, N; jj = j, \dots, \frac{N}{2}; ii = i, \dots, \frac{M}{2}$.

In Eqs. (3)–(6), the following notation is used: LL1 [ii, jj]—an array of approximation coefficients; HL1 [ii, jj]—an array of horizontal detail coefficients; LH1 [ii, jj]—an array of vertical detail coefficients; and HH1 [ii, jj]—an array of diagonal detail coefficients.

The arrays of coefficients of approximation and detailing of level 1 obtained using 2D DVP have the same dimension [ii, jj], which means that each of these arrays is two times smaller than the array of the original image.

To get the arrays of coefficients for the second level of decomposition: LL2, HL2, LH2, HH2, you need to perform similar actions, while taking the array of approximation coefficients LL1 obtained at the first level of decomposition for the original image.

At each decomposition level, coefficients exceeding the threshold values are selected. The threshold T_i for the decomposition of level i depends on the maximum absolute coefficient C_i over all ranges of coefficients of level i, thus [14]:

$$T_i = 2^{\left[\log_2 C_i\right]-1} \tag{7}$$

The watermark is embedded in all coefficients exceeding the threshold value T_i for the corresponding decomposition level obtained from Eq. (7). Used additive algorithm for embedding watermark:

$$v_i'(x,y) = v_i(x,y) + \alpha * v_i(x,y) * x_i(x,y) \tag{8}$$

where $v_i'(x,y)$ is a modified coefficient with coordinates (x, y); $v_i(x,y)$ is a coefficient chosen for implementation with coordinates (x, y), for example, for level 1: LH1 (x, y), HH1 (x, y), HL1 (x, y)); $x_i(x,y)$—bit of watermark with coordinates (x, y); α—the scaling factor is adjusted for each level of decomposition. So for the range of coefficients of approximation of the third level α = 0.02, since the coefficients of this range have large values. Scaling factors of 0.1, 0.2, and 0.4 are used for the third, second, and first levels of decomposition, respectively. The fractal image is first embedded in the wavelet coefficients of the third level, then, if the coefficients of the third level were not enough to fully embed the watermark, the coefficients of the second level are used, etc. To extract the watermark, the inverse of the implementation equation (Eq. (9)) is used, but the adaptive-level scale factor α is taken into account.

$$x_i'(x,y) = \frac{v_i'(x,y) - v_i(x,y)}{\alpha * v_i(x,y)} \tag{9}$$

A fractal image, in which a watermark is embedded, acts as a watermark.

4. Experimental results

Figure 2a shows the original unfilled image with a size of 6400 × 4096 pixels in JPEG format. The image containing the embedded key is shown in **Figure 2b**.

Figure 3 shows the results of extracting the secret fractal key from the filled image. **Figure 3a** shows the original fractal image, in **Figure 3b** is the image obtained after extraction.

a) b)

Figure 2.
(a) Original cover image; (b) stegoimage with embedded fractal key.

a) b)

Figure 3.
(a) Generated fractal with watermark; (b) extracted from stegoimage fractal image with watermark.

a) b)

Figure 4.
(a) Original watermark; (b) watermark extracted from fractal key.

Figure 4 shows the results of extracting a digital watermark from the selected fractal. **Figure 4a** contains the original watermark and **Figure 4b** shows watermark extracted from the key.

As a result of the algorithm, a watermark was extracted with high quality. Despite the fact that you can observe small pixel distortions in the extracted secret key, the received watermark as a QR code is completely identical to the original one.

5. Evaluation of the quality of the algorithm

The quality assessment of the visual distortion of the fractal cover image was performed on the basis of the following metrics: normalized mean square error (NMSE) and peak signal-to-noise ratio (PSNR) (Eqs. (10) and (11)):

$$NMSE = \sum_{x,y} \frac{\left(C_{x,y} - S_{x,y}\right)^2}{\sum_{x,y}\left(C_{x,y}\right)^2} \tag{10}$$

$$PSNR = XY \cdot \max{}_{x,y} \frac{\left(C_{x,y}\right)^2}{\sum_{x,y}\left(C_{x,y} - S_{x,y}\right)^2} \tag{11}$$

In the relations presented, $C_{x,y}$ denotes a pixel with the coordinates (x, y) of the empty cover image, and $S_{x,y}$ denotes the corresponding pixel of the filled image.

The calculation of the metrics was carried out with different sizes of the implemented secret key and digital watermark. The results are presented in **Table 1**.

To assess the secrecy of the developed system, a situation was simulated in which the attacker took possession of a secret key with an embedded watermark, and he also knows some parameters of the generated fractal, namely, the size of the rectangle l, the maximum number of iterations k and the image size m. The only unknown parameter is the starting point c. As an example, four fractals were generated with different values of the parameter c (**Figure 5**) and the watermark was extracted using the obtained images and the secret key containing the watermark presented in **Figure 4b**.

The results of the extraction of watermark are shown in **Figure 6**.

As can be seen from the figures, all the images obtained as a result of the experiment are so significantly different from the original OT. As a result, the extraction of information from the QR code becomes impossible. From the experiment, it is clear that to obtain the information it is necessary to know the parameters of the fractal key with high accuracy. The developed method makes it possible to reduce the probability of the substitution or theft of secret information in similar steganographic systems to almost zero.

Size of watermark, pixels	Fractal key size, pixels	NMSE	PSNR
30 × 30	85 × 85	0.0021	31,1922
35 × 35	100 × 100	0.0031	29,4217
40 × 40	115 × 115	0.0056	26,7753
45 × 45	130 × 130	0.006	26,5334
50 × 50	150 × 150	0.0067	26,3538

Table 1.
Values of NMSE and PSNR with different sizes of embedded data.

DOI: http://dx.doi.org/10.5772/intechopen.92018

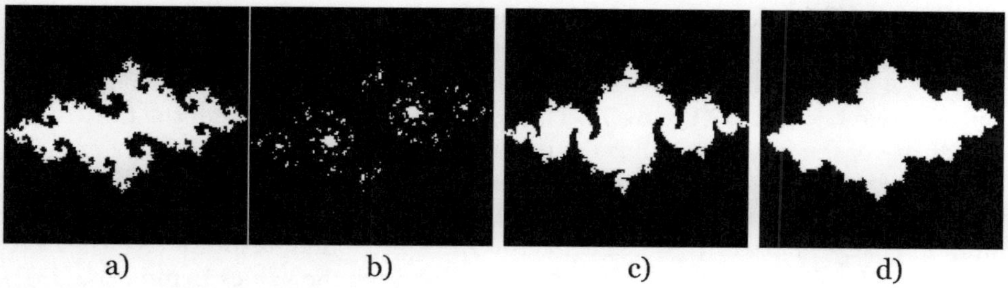

Figure 5.
*Fractal images with different starting point c. (a) −0.73949 + 0.16498*i; (b) −0.74549 + 0.37841*i; (c) −0.80939 + 0.12388*i; (d) −0.63949 + 0.19098*i.*

Figure 6.
Watermarks obtained during extraction using fractals shown in ***Figure*** *5.*

6. Analysis of the possibility of steganalysis by measuring the fractal dimension of the secret key before and after embedding

The importance of steganographic watermark embedding is their steganographic attacks resistance. In this case, the problem arises of detecting the very fact of the introduction of the watermark into the cover image. For this purpose, the method of estimating the fractal dimension of the image before and after embedding the watermark can be used. Consider the problem of measuring the fractal dimension of color or black and white images after embedding the watermark into them [15].

As is known, a fractal is defined as a collection of objects for which the Hausdorff dimension is strictly greater than the topological dimension. The concept of self-similarity is used to estimate the fractal dimension.

A bounded fractal set in a Euclidean n-space will be self-similar if it is a union of different N (disjoint) reduced copies that can be scaled using a special scaling factor r. In accordance with the entered scaling factor, the fractal dimension D of set A can be obtained using Eq. (12).

$$D = \frac{\log(N)}{\log(1/r)} \tag{12}$$

where N is the total number of boxes L needed to cover the fractal set; 1/r is the scaling factor of the box in relation to the image.

As a result, D is a dimension relative to the size of the box used to measure the fractal image. It can be said that the fractal dimension is a measure of how "complex" a self-similar figure is.

Let us consider the two most common methods for measuring dimension: differential box-counting and triangulation method.

The differential box-counting (DBC) method takes into account the difference between the maximum and minimum intensity values of the brightness of the image. Let an image be given with the size M × M, which is divided into "boxes" with the size L × L × L′ (**Figure 7**). The height of the box L′ divides the third coordinate of the image, which is the intensity value.

Space (x, y) of the image containing the values of the coordinates of the peak-mudflows is divided into cells of size L × L, after which the maximum and minimum intensity values of (i, j) cell equal to l and k are found, respectively. The next step for each cell is the sum of the differences between the values found (Eq. (13)).

$$n_{r(i,j)} = (l - k + 1)/L' \tag{13}$$

where r = L/M is the reduction factor.

After calculating the amount in all cells is the total amount of differences for the entire image (Eq. (14)).

$$N_r = \sum_{i,j} n_r(i,j) \tag{14}$$

The regression curve of log dependence $\log\left(\sum N_r\right)$ on log (1/R) is constructed on the basis of the calculations. Fractal dimension D is defined as the tangent of the slope angle of the curve.

When using the triangulation method (TM), the image is divided into identical cells ε of size s × s. Four heights equal to the intensities of pixels in the corners of the cells (a–d) are considered. At the intersection of the diagonals of the cell, the point (e) is set, the value of which is equal to the arithmetic average of four heights. If we represent a cell in the form of a triangular prism, as shown in **Figure 9**, then it is necessary to calculate the area of the projected upper surface, shown in **Figure 8**.

First of all, the values of the sides of the four triangles, obtained by connecting the diagonals of the cell, are calculated (Eqs. (15) and (16)).

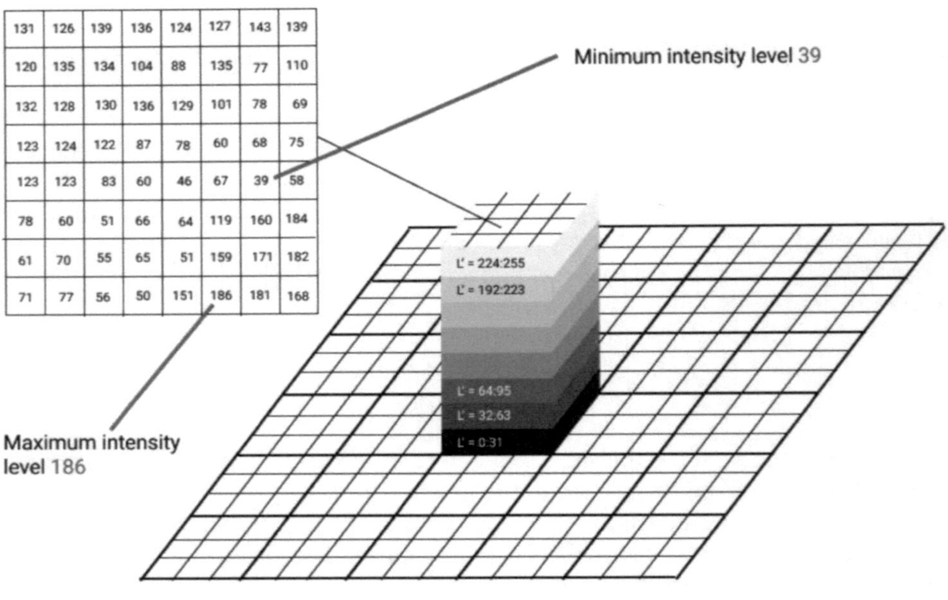

Figure 7.
Three-dimensional representation of the image with its gray levels [15].

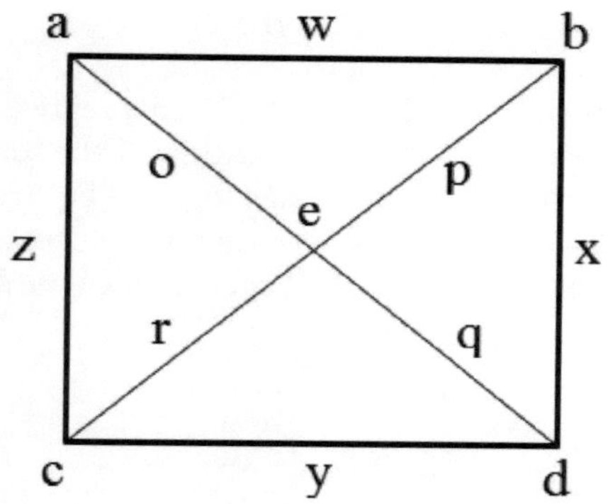

Figure 8.
The projected top surface of a triangular prism.

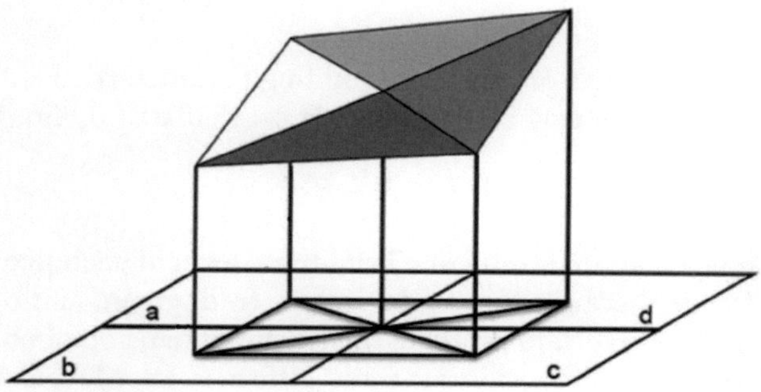

Figure 9.
Representation of a cell of the image in the form of a triangular prism.

$$w = \sqrt{(b-a)^2 + s^2}; x = \sqrt{(c-b)^2 + s^2};$$
$$y = \sqrt{(d-c)^2 + s^2}; z = \sqrt{(a-d)^2 + s^2}; \tag{15}$$

$$o = \sqrt{(a-e)^2 + \left(\sqrt{2}/2s\right)^2}; p = \sqrt{(b-e)^2 + \left(\sqrt{2}/2s\right)^2};$$
$$r = \sqrt{(c-e)^2 + \left(\sqrt{2}/2s\right)^2}; q = \sqrt{(d-e)^2 + \left(\sqrt{2}/2s\right)^2}. \tag{16}$$

Then, using the Heron formula, the semi-perimeters and areas of all triangles are calculated (Eqs. (17) and (18)).

$$A = \sqrt{sa(sa-w)(sa-p)(sa-o)}; B = \sqrt{sb(sb-x)(sb-p)(sb-q)};$$
$$C = \sqrt{sc(sc-y)(sc-q)(sc-r)}; D = \sqrt{sd(sd-z)(sd-o)(sd-r)}. \tag{17}$$

$$sa = \frac{1}{2}(w+p+o); sb = \frac{1}{2}(x+p+o);$$
$$sc = \frac{1}{2}(y+q+r); sd = \frac{1}{2}(z+o+r). \tag{18}$$

The total surface area is equal to the sum of the areas of individual triangles (Eq. (19)).

$$S_{ABCD} = A + B + C + D \tag{19}$$

This procedure is repeated for all cell sizes. Then a regression line is constructed, which determines the dependence of the logarithm of the total area of all the triangles log (S) on the logarithm of the cell size log (ε) (**Figure 10**).

To calculate the fractal dimension, it is necessary to find the tangent of the angle of inclination of the constructed curve B. It can be calculated using Eq. (20).

$$B = \frac{r * S_s}{S_\varepsilon}$$

$$r = \frac{\text{cov}(\varepsilon, S)}{S_\varepsilon S_S}; \text{cov}(\varepsilon, S) = \frac{\sum(\varepsilon_i - \overline{\varepsilon})(S_i - \overline{S})}{N};$$

$$S_\varepsilon = \sqrt{\frac{\sum(\varepsilon_i - \overline{\varepsilon})}{N}}; S_S = \sqrt{\frac{\sum(S_i - \overline{S})}{N}}. \tag{20}$$

$\overline{\varepsilon}, \overline{S}$ are the average values of the corresponding parameters.

The desired value of the fractal dimension D is calculated by Eq. (21).

$$D = 2 - B \tag{21}$$

In grayscale images, there is only one brightness level of each pixel of the image, while in a color image there are three color values (red, green, and blue) for each pixel of the image. As a result, to estimate the fractal dimension of color images, it is necessary to calculate the dimension for three different color levels. Using the estimation methods of calculating the fractal dimension for images in grayscale, one can calculate the dimension of a color image.

To calculate the dimension of a color image using the DBC method, the same method of calculating dimensions is used as in the grayscale images. In the color image, the dimension is calculated for the red R, green G, and blue B components. According to the DBC technique for each stage (red, green, and blue), you can apply the technique to images in shades of gray. After finding the results of each step, the results of different color levels are combined. As a result, you can get the dimension of the color image:

R: $n_{rr}(i,j)$ = l-k + 1 for red values;
G: $n_{rg}(i,j)$ = l-k + 1 for green values;
B: $n_{rb}(i,j)$ = l-k + 1 for blue values.

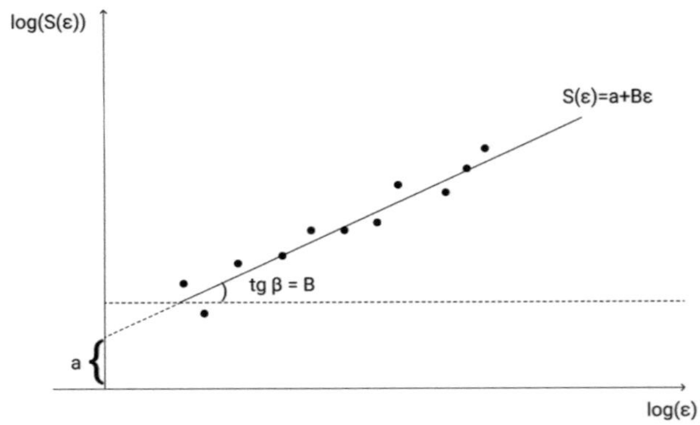

Figure 10.
Regression line.

According to the above, you can find the fractal dimension of the color image $N_{r\{r,g,b\}}$ (Eq. (22)).

$$N_{r\{r,g,b\}} = \frac{\sum_{i,j} n_{rr}(i,j) + \sum_{i,j} n_{rg}(i,j) + \sum_{i,j} n_{rb}(i,j)}{3} \tag{22}$$

When calculating the dimension of color images by the triangulation method, first of all, the particular values of the fractal dimension for each color component are determined, and then their average value is calculated.

7. Testing methods for estimating fractal dimension

To test the methods for evaluating the fractal dimension, two methods triangulation and DBC were considered. Testing was carried out with the help of fractal objects for which the dimension is known. **Table 2** presents the results of testing the above assessment methods using fractals with a known dimension value.

In **Table 2**, the following designations are made:

$\Delta_{TM}\% = \frac{\hat{D}_{TM} - D}{D} \cdot 100\%$ is the absolute error in estimating the dimension between the measured and the true value of the fractal dimension by the triangulation method.

$\Delta_{DBC}\% = \frac{\hat{D}_{DBC} - D}{D} \cdot 100\%$ is the absolute error in estimating the dimension between the measured and the true value of the fractal dimension by the DBC method.

$\Delta_{AV}\% = \frac{\overline{D} - D}{D} \cdot 100\%$ is the absolute error of the average estimate of the fractal dimension obtained by two methods.

Name of fractal	Illustration	D	Testing results					
			TM	Δ_{TM},%	DBC	Δ_{DBC},%	$\overline{D}$	Δ_{AV},%
Dodecahedron fractal	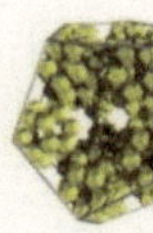	2.3296	2.2276	4.378	2.4203	3.893	2.3239	0.243
Fractal pyramid		2.3219	2.2641	2.489	2.5524	9.03	2.4083	3.719
Jerusalem cube		2.529	2.3218	8.193	2.5830	2.135	2.4524	3.029
3D quadratic Koch surface		2.3347	2.2260	4.656	2.4068	3.088	2.3164	0.784
3D quadratic Koch surface (type 2)		2.5	2.3879	4.484	2.5705	2.82	2.4792	0.832

Table 2.
Results of testing methods for measuring the fractal dimension.

The numerical data of the absolute errors of measurement of the dimension with the help of triangulation and DBC methods show that they give comparable results. The error of calculations for both methods does not exceed 5%. For practical use, it is advisable to use both methods, and the dimension values to find by averaging the results of both methods. In this case, the total error of calculations does not exceed 3%.

8. The measurement results of the fractal dimension of fractal keys

To illustrate the possibility of detecting an embedded watermark in images, we consider the experimental results of the evaluation of the fractal dimension after embedding the watermark in a fractal image based on Julia set.

Table 3 shows the results of measuring the fractal dimension of color fractal images in the form of a Julia set when embedding watermark by various known steganographic methods.

The results show that the steganographic introduction of data does not actually affect the value of the fractal dimension, which makes it impossible to illegally extract the watermark.

Starting point *c*	Number of iterations	Size of rectangle	Embedding method	Image size	D of original fractal	D of fractal with watermark
−0.7778+ 0.1316i	300	1.5	LSB	1024	2.318	2.318
			Block based		2.318	2.317
			Koch R		2.318	2.316
			Koch G		2.318	2.316
			Koch B		2.318	2.318
−0.74543 +0.11301i			LSB		2.438	2.438
			Block based		2.438	2.437
			Koch R		2.438	2.436
			Koch G		2.438	2.436
			Koch B		2.438	2.438
	3000		LSB		2.109	2.109
			Block based		2.109	2.115
			Koch R		2.109	2.136
			Koch G		2.109	2.097
			Koch B		2.109	2.112
		0.5	LSB		2.192	2.192
			Block based		2.192	2.194
			Koch R		2.192	2.208
			Koch G		2.192	2.178
			Koch B		2.192	2.195
		0.0005	LSB		2.172	2.172
			Block based		2.172	2.179
			Koch R		2.172	2.186

Starting point *c*	Number of iterations	Size of rectangle	Embedding method	Image size	D of original fractal	D of fractal with watermark
			Koch G		2.172	2.165
			Koch B		2.172	2.359
	300	1.5	LSB	512	2.449	2.449
			Block based		2.449	2.442
			Koch R		2.449	2.427
			Koch G		2.449	2.425
			Koch B		2.449	2.450
		0.5	LSB		2.541	2.541
			Block based		2.541	2.546
			Koch R		2.541	2.523
			Koch G		2.541	2.523
			Koch B		2.541	2.537
		0.0005	LSB		2.396	2.396
			Block based		2.396	2.383
			Koch R		2.396	2.474
			Koch G		2.396	2.423
			Koch B		2.396	2.487
	3000		LSB		2.183	2.183
			Block based		2.183	2.183
			Koch R		2.183	2.221
			Koch G		2.183	2.323
			Koch B		2.183	2.122

Table 3.
Changing the value of the fractal dimension of images during steganographic embedding.

9. Conclusions

The secrecy of steganographic systems is based on the assumption that the attacker is not aware of the fact of the information being introduced. In the event that this fact becomes publicly available, it will not be difficult for an unauthorized user to extract secret data or delete a given watermark. To solve this problem, it was proposed to use an additional container in the form of an algebraic fractal in the form of a Julia set.

Fractal generation is carried out using predefined secret parameters using the escape time algorithm (Escape time algorithm).

Embedding a digital watermark in a container was carried out in two stages. At the first stage, the watermark is added to the generated fractal. The resulting image is embedded in the original container in JPEG format. As a result, the original image has almost no visual distortion. The measurement of the values of the NMSE and PSNR metrics confirmed the high level of embedding quality and extraction of the watermark in the form of a QR code.

To confirm the high level of secrecy, an experiment was conducted, during which an attempt was made to replace the original secret key. The experiment confirmed that it is impossible to extract the watermark without knowing the true parameters of the fractals.

To assess the possibility of detecting the fact of embedding a watermark on the basis of fractals, the fractal dimension of intermediate containers was measured on the basis of the Julia set. Measurement of the dimension value was carried out before and after steganographic embedding. The results showed that the fractal dimension varies slightly, within the limits of the method error, and cannot be a sign that characterizes the presence of an integrated watermark.

The proposed technique allows for steganographic embedding with a high level of visual quality, as well as resistance to various steganographic attacks. As a result, it is possible to increase the level of secrecy of the watermark embedding algorithms and significantly reduce the likelihood of data theft.

Author details

Oleg Sheluhin and Dzhennet Magomedova*
Information Security Department, Moscow Technical University of Communication and Informatics, Moscow, Russia

*Address all correspondence to: jimagomedova@gmail.com

References

[1] Thamizhchelvy K, Geetha G. Design of digital signature algorithm by fractals and chaos theory. International Journal of Computer Applications. 2012;**5**:50-57. DOI: 10.5120/4608-6588

[2] Aslan N, Saltan M, Demir B. A different construction of the classical fractals via the escape time algorithm. Journal of Abstract and Computational Mathematics. 2018;**4**:1-15

[3] Rani M. Cubic superior Julia sets. In: Proceedings of the European Computing Conference; 2011. pp. 80-84. ISBN: 978-960-474-297-4

[4] Xing Y, Tan J, Hong P. Quaternion Julia fractals. In: IEEE Computer Society, The 9th International Conference for Young Computer Scientists; 2008. pp. 797-802

[5] Nori AS, Al-Qassab AM. Steganographic technique using fractal image. International Journal of Information Technology and Business Management. 2014:52-59. ISSN: 2304-0777

[6] Sheluhin OI, Kanaev SD. Algorithms and software implementation. In: Sheluhin OI, editor. Steganography. Moscow: Goryachaya liniya – Telekom; 2017. p. 616. ISBN: 978-5-9912-0579-5

[7] Sheluhin OI, Smychek MA. Combined graphic scale image watermarking method for secret communication. Radiotehnicheskie i Telekommu-Nikatsionnyie Sistem. 2017;**1**:68-75. ISSN: 2221-2574

[8] Abbas TA, Hamza HK. Steganography using fractal images technique. IOSR Journal of Engineering (IOSRJEN). 2014;**4**:52-61. ISSN(e): 2250-3021

[9] Thamizhchelvy K, Geetha G. Data hiding technique with fractal image generation method using chaos theory and watermarking. Indian Journal of Science and Technology. 2014;**7**: 1271-1278. ISSN(Online): 0974-5645

[10] Hardikkumar Desai V, Desai Apurva A. Image Steganography Using Mandelbrot Fractal. International Journal of Computer Science Engineering and Information Technology Research (IJCSEITR). 2014; **4**:71-80. ISSN(E): 2249-7943

[11] Wu Y, Noonan J. Image steganography scheme using chaos and fractals with the wavelet transform. International Journal of Innovation, Management and Technology. 2012;**3**: 285-289

[12] Sheluhin OI, Kanaev SD. Hiding watermarks in color images using algebraic fractals by 2D wavelet transform methods. T-Comm. 2018;**6**: 46-50

[13] Xia X, Boncelet C, Arce G. Wavelet transform based watermark for digital images. Optics Express. 1998;**12**:497-511. DOI: 10.1364/oe.3.000497

[14] Sheluhin OI, Magomedova DI. Analysis of methods for calculating the fractal dimension of color and gray-scale images. H&ES Research. 2017;**6**:6-16. ISSN: 2412-1363 (e)

[15] Padhy LN. Fractal dimension of gray scale & colour image. International Journal of Advanced Research in Computer Science and Software Engineering. 2015;**7**:1285-1297. ISSN: 2277 128X

Chapter 6

Fractal Analysis of Strain-Induced Microstructures in Metals

Ricardo Fernández, Gaspar González-Doncel and Gerardo Garcés

Abstract

The deformation of materials is a key topic for different industrial sectors. The correlation between specific thermomechanical processes, like extrusion, rolling or additive manufacturing, and the resultant material's microstructure, is particularly interesting. In these thermomechanical processes, the microstructure of the materials depends mainly on the applied stress, the magnitude of strain achieved in a given time period, and the temperature. In the case of metals and alloys, plastic deformation can be described microscopically based on the dynamics of a huge population of moving dislocations. Plasticity is characterized at the mesoscale by the avalanche-like collective behavior of dislocations, which is a typical case of self-organized systems. Dislocations are organized into cells named grains or subgrains that greatly influence the mechanical behavior of metals and alloys. The existence of these fractal structures of dislocations in metals is well established. However, it is very complex to conduct a fractal analysis of these microstructures. This is actually done by looking for a compromise between a detailed description of the complex fractal microstructure and the development of a practical procedure that avoids unaffordable extensive characterization in a different time and spatial scales. Several cases will be described considering different alloys and experimental conditions.

Keywords: fractal, metal, strain, dislocations, mesoscale

1. Introduction

The concept of fractal refers to structures whose appearance is independent of the scale at which it is observed. In these geometrical objects, the same pattern appears regardless of whether we move away to have a more global vision of the observed structure or if we approach to expand a particular detail of it. Fractals are very common in many aspects of nature. They appear, for example, in the crystalline structure that develops in snowflakes or in the crystals that form frost during cold and wet sunrises. The rough, superficial structure of broccoli is another recurring example in living beings where a fractal nature is revealed. The structure of the veins through which the sage passes on the leaves of the trees is another very clear example of a fractal. The coastline can be described, in some cases, in the frame of a fractal object. Fractals appear not only in nature and are not restricted to physical objects, but they are also present in other fields, such in financial markets; the repeated and cyclic behavior of the stock market also falls in a fractal patter. Music is other field in which fractals have been reported, for example, when shorter passages are reflected in expanded form in longer passages. It is surprising that, being fractals so common objects and present in so many fields, the concept of fractal was initially proposed as late as in 1975 [1]. This "self-similarity" of fractal geometry was proposed by the IBM mathematician Benoît Mandelbrot (the term fractal derives from the Latin word *fractus*).

The widespread appearance of the concept of fractal in many areas is also observed in different microstructural features of materials. In fact, the appearance of fractal structures in metals is widely accepted [2]. Solid state metals have an orderly crystalline structure that provides their physical properties. The strongest metal bond is formed when the metallic atoms are ordered as close as possible, usually in cubic or hexagonal lattices. These lattices are described by the unit elemental cell characterized by the lattice parameter, dimension measured in nm and their symmetry typically represented by cubic or hexagonal unit cells. Most of the metals we find in our daily lives are polycrystals. This means that not all the unit elemental cells of the material have got the same orientation with respect each other. Actually, there are domains of atoms/unit cells where their orientation is the same. These domains, which are present in millions in a metallic component, are called grains. These grains are separated from each other by grain boundaries. All this microstructural complexity is a fertile breeding ground for fractal structure appearance. One example is the presence of fractals in the solidification/crystallization structure generated in the manufacturing process of polycrystalline metals during cooling [3]. Another example of fractal structures in metals is provided by the distribution of reinforcement in discontinuously reinforced composite materials. The physical properties of these composite materials are governed by the properties of the clusters that are formed when contacts are established among the reinforcing particles. The percolation of the system greatly modifies its properties. The formation of these clusters depends on the fractional content of the composite material [4]. Another classic example of fractal structures in metals is the fracture surface [5]. These fracture surfaces are generated from the interaction of macroscopic defects, mainly cracks due to the application of mechanical stresses that generate a progressive deformation of the material until its breakage. These surfaces are formed by facets of different sizes and heights that can be accurately described by a fractal. The knowledge of the fractal nature of fracture surfaces has allowed the design of fractal surfaces to achieve certain properties such as a better air flow avoiding turbulence [6].

The fractal nature of the fracture surfaces is an indication that the deformation process generates fractal structures in the deformed microstructure, which ultimately give rise to the fractal fracture surface. In the case of metals, the deformation process begins with an elastic deformation that occurs in small intervals typically smaller well below 0.2% strain. The stress value achieved at this point is known as the yield stress, and it has been agreed that it reflects the beginning of plastic deformation of metals in the macroscale (homogeneous deformation). From this strain/stress value, the deformation is mainly caused by the movement of dislocations. The main defects in metals are vacancies and dislocations. These defects correspond to the lack of a semi-plane of atoms in its crystalline structure, respectively. When the temperature is high and the applied stress is low, this movement can be controlled by the motion of vacancies (diffusion). Dislocations are found not only inside the polycrystalline grains but also at the grain boundaries. A characteristic of dislocations, the fundamental element in metal deformation, is that their collective arrangement has a fractal nature. Sevillano et al. [7] were the first to recognize the possibility of fractal geometry of the cell structure based on the measurement of the fractal dimension of the microstructure of cold worked Cu. These dislocations are organized during the deformation process and form cell structures that have a fractal nature [8]. Moreover, the tangled arrangement of dislocations in metals has been proposed to be associated with a fractal structure. Examples are found in various usual metal forming processes such as rolling [9].

It is important to understand the role of the collective motion and interaction of a large number of dislocations in the generation of fractal structures [10]. The sample size is one of the features that determine the fractal dimension of these dislocation structures [11]. In that study, carried out in iron micropillars, it has been observed that the fractal dimension of the dislocation structure changes approximately 40%, from 1.1 to 1.5, when the sample diameter increases from 300 to 1500 nm. This reveals the strong size effect on fractal substructures of dislocations in metals.

2. Origin of fractal structures in metals: stochastic vs. thermodynamic approach

Materials, in particular metals, can be considered complex systems from the mechanical point of view, considering that they contain a large number of elements and dislocations that interact with each other. A complex system has properties that arise from the interaction between its elements and the system itself with the environment. In particular, one of these properties is the emergency. The emergency can be defined as the generation of spontaneous patterns resulting from the self-organization of the elements of the system. Classic examples of an emergency in nature are snowflakes. These patterns have different geometries with a high degree of symmetry indicating the ability to organize the system. Surprisingly, a similar behavior has been found in the case of gallium-based liquid metals. By applying an electric potential to this liquid metal, snowflake-shaped fractals have been found [12]. These structures have encouraged the introduction of the term fractalized metals. It has been proposed that this concept defines a new state in which metals can be found. In this state, all positions in the metal lattice are coordinated granting properties such as chiral transport [13].

Metal microstructures have a great amount of defects. Dislocations, linear defects, are the most important ones from the point of view of plastic deformation and mechanical behavior. An individual, isolated, dislocation has an associated elastic stress field. For an edge dislocation, is described, in Cartesian coordinates, by a tensor where the components, σ_{ij}, are given by Eq. (1) [14].

$$\sigma_{xx} = -\frac{Gby(3x^2+y^2)}{2\pi(1-\nu)(x^2+y^2)^2};\sigma_{yy} = -\frac{Gby(x^2+y^2)}{2\pi(1-\nu)(x^2+y^2)^2};\sigma_{xy} = \sigma_{yx} = \frac{Gbx(x^2-y^2)}{2\pi(1-\nu)(x^2+y^2)^2};\sigma_{zz} = \nu(\sigma_{xx}+\sigma_{yy});\sigma_{xz} = \sigma_{zx} = \sigma_{zy} = \sigma_{yz} = 0 \quad (1)$$

where G is the shear modulus, b is the Burgers vector, ν is Poisson's ratio, and x, y, and z are the coordinates, perpendicular in-plane, perpendicular out-of-plane, and parallel directions, with respect to the dislocation line, respectively, in the reference system of the dislocation.

The specific stress field depends on the character of the dislocation, i.e., screw or edge. The stress fields of a huge amount of dislocations, in combination with the external applied stress, play a relevant role in the formation of fractal deformation patterns in metals. Through these stress fields, interactions occur with other dislocations or with second phases of the material, such as precipitates. The interaction among typically 10^{10}–10^{14} m^{-2} dislocations generates fractal patterns in the mesoscale. The formation of these fractals follows the same fundamental rules as any other physical system which develops a fractal structure. Here, it is particularized to the specific characteristics of metals. The formation of fractals in physical systems can be explained from a stochastic or a thermodynamic point of view. The concept that represents the susceptibility of uniform systems to spontaneously develop patterns in uniform nonlinear systems is known as Turing instability [15]. These patterns are found in reaction–diffusion systems describing, for example, the transformation of substances by chemical reactions in which the substances are transformed among each other and spread out over by diffusion. In the case of metals, a reaction–diffusion model of coupled nonlinear equations has been used to describe the formation of forest (immobile) and gliding (mobile) dislocation densities in the presence of cyclic loading [16]. This description is done assuming that a metal undergoing both an internal and external stress is far from equilibrium. The second law of thermodynamics, states that any physical system can spontaneously increases its entropy, i.e., from order to disorder. The apparent conflict between this postulate and the generation of ordered fractal patterns is dismissed by considering that the accompanying decrease in entropy (i.e., ordering) is compensated by a corresponding increase somewhere else, so that the net entropy always increases. However, very recent theories [17] postulate that the apparent conflict between the second law of thermodynamics and ordered systems evolution is due to the consideration that the Universe is near equilibrium. This new theory postulates that there is an energy/matter flow comprising the whole Universe that evolves from disorder to order via self-organization. This process must be described by the empirical laws of non-equilibrium thermodynamics [18]. In this situation, a stress gradient in relatively simple non-equilibrium systems, as deformed metals, causes a flux of energy/matter in the system. As a consequence, it emerges a countervailing gradient. These conjugated processes result in spatio-temporal macroscopic order that spontaneously emerges provided that system is driven far away enough from equilibrium [19, 20]. Some authors have recently accepted that the contribution of the macroscopic (or mesoscopic) order, described by the entropy, must be considered for a rigorous description of the mechanical behavior of metals at high temperature [21].

Stochastics models have been also used to justify the appearance of fractal patterns developed during strain of metals. The necessity of a stochastic treatment, specific for fractal structures, for the characterization of mesoscopic dislocation network distributions in metals was established at the very end of the twentieth century [22]. At temperatures below 0.5 T_{melt} (with T_{melt}, the melting temperature), it is considered that there is no time-dependent plasticity (creep). In this range of temperatures, plasticity is explained by the avalanche-like collective behavior of dislocations [23]. The self-organized structures of dislocations generated by this phenomenon are developed at the grain size scale. This can be described as the mesoscale in metals. Natural nondeterministic fractals are self-similar in a statistical sense over a wide range of scales [1]. The localization of the plastic deformation of metals and alloys into slip bands has been investigated for a long time [24]. One example of dislocation patterns developed under uniaxial tensile tests appears with the Lüders band formation in some steels. The appearance of these patterns under uniaxial stress conditions reveals the great importance of dislocations interaction in their formation.

3. Experimental characterization of fractal structures

There are different parameters that can be used to describe a fractal. The most common one is the fractal dimension. Fractal dimensionality is considered a measure of complexity for systems that do not present integer dimensions [25]. However, there are different fractals that have the same fractal dimension. Therefore, other parameters have been proposed to describe some characteristics of fractals. One of the most important characteristics of fractal structures of dislocations generated during metal deformation is their connectivity. Dislocations are usually arranged in grain and subgrain boundaries. These boundaries maintain their integrity, while the deformation increases due to the stress fields of dislocations that lock them in low-energy configurations. The connectivity of these structures can be described in this context by the distance to next dislocation neighbors. In particular, lacunarity is one of the most used parameters to characterize the connectivity among different parts of a fractal structure. Lacunarity is a measure of spatial heterogeneity, and it is used to differentiate images with similar fractal dimension but different appearances [26]. The local connected fractal dimension represents the connectivity of the different parts of the system by a color map that indicates the proximity of the neighbor pixels in a given image.

3.1 Image analysis

All parameters that describe a fractal structure of dislocations can be calculated from images obtained from the microstructure. In the deformation of metals, the most interesting images are related with the observation of dislocations that can be homogeneously distributed or rearranged to form subgrains or grain boundaries. These dislocation structures have been described in a very detailed manner from decades thanks to transmission electron microscopy (TEM). However, there are some aspects that must be taken into account when analyzing the fractal parameters of dislocation microstructures. The first one is the quality of the images. Normally, metallographical images have a gray range. In the case of a TEM image, moreover, there is a great contrast evolution depending on the interaction between the electrons and defects (dislocations, stacking faults, etc.), the relative orientation of the crystal, and the mode in which the images are obtained (bright field vs. dark field).

Figure 1 shows a TEM bright field image of the microstructure of a magnesium alloy which was deformed at intermediate temperatures. The presence of dislocations is clearly observed. The fractal dimension is typically measured from this kind of images using the box counting method.

The "box counting" is a method for analyzing complex patterns by splitting an image into smaller and smaller pieces, typically "box"-shaped. The fractal dimension is calculated considering the number of boxes containing part of the microstructure vs. the number of empty boxes. The ratio of full vs. empty boxes is considered for at least 4/5 box sizes for the analysis of the fractal dimension. Computer-based box counting algorithms have been applied to patterns in one-, two-, and three-dimensional spaces [27, 28]. The technique is usually implemented in specialized software to analyze patterns extracted from digital images. This software also has application in the determination of related parameters such as lacunarity and multifractal analysis [29].

The TEM photos must be treated to obtain a binary image. However, in this process, TEM images cannot be completely clean of artifacts such as shadows or dots. The presence of these artifacts introduces an overestimation of the fractal dimension.

Figure 2 shows several idealized examples presenting this situation. **Figure 2(a)** shows an idealized grain before deformation. Therefore, no dislocations or subgrain boundaries appear. After deformation, we can calculate the fractal dimension, D_F, of, as an exercise, one or two dislocations within this grain (**Figure 2(b)** and **(c)**, respectively). For these two ideal examples, the fractal dimension calculated with

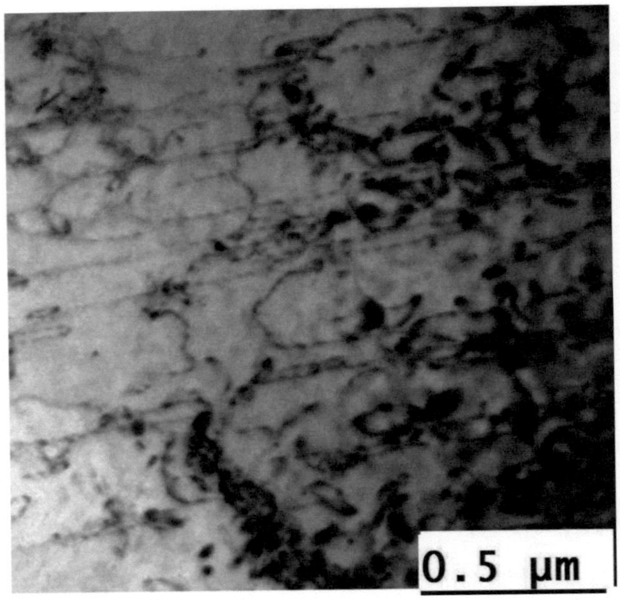

Figure 1.
Bright field image of a magnesium alloy deformed around 2% in tensile mode at a temperature of 200°C and a strain rate of 10^{-3} s^{-1}. Zone axis $[11\bar{2}0]$ g = $[1\bar{1}00]$.

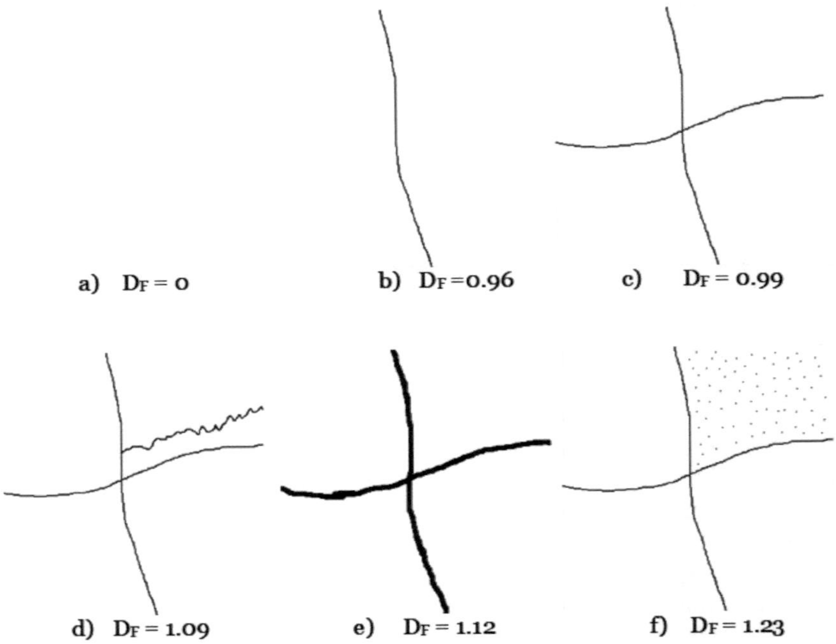

Figure 2.
(a) Idealized microstructure of a completely dislocation-free grain before plastic deformation. (b–d) Subgrain formation during the deformation process. (e) Subgrain coarsening during deformation process. (f) Microstructure shown in (c) with an artifact.

the plugin FracLac [30] used in FIJI software is very similar (0.96 and 0.99, respectively). Since dislocations are line defects, these values are very reasonable. As the macroscopic strain increases, new dislocations are generated, or, even, subgrains can be developed. This increase in dislocation density results in an additional increase in the fractal dimension (**Figure 2(d)** and **(e)**). As it is shown in **Figure 2 (b)–(e)**, the fractal dimension of dislocation substructures is related to the area covered by dislocations/substructure. Then, the fractal dimension of the structure shown in **Figure 2(e)** formed by thick lines is 10% greater than the structure shown in **Figure 2(b)**. However, there are some elements, mainly dots, which greatly influence the fractal dimension calculated by the box counting technique. As an example, **Figure 2(f)** shows the same structure as **Figure 2(c)**, as far as it is assumed that during the binarization process, the dusty dots are part of the image. These dusty dots are considered as an artifact. As a result of the consideration of the dusty points in the image, the fractal dimension increases around 25%.

There are other aspects that must be taken into account when the fractal dimension of a dislocation structure in a deformed metal is calculated using TEM images. The first one is related to the fact that TEM images of the dislocations can be taken under several 3D orientations of the crystal and/or using different diffraction vectors. Dislocations can be visible only when the scalar product g × b (where g is the diffraction vector of the crystal that is excited, i.e., Bragg's conditions are attained) is different from zero. **Figure 3** shows an example of a magnesium alloy deformed at room temperature. The Burgers vector in magnesium is [$11\bar{2}0$]. Dislocations are invisible when g = [0002] (a) is excited and visible with g = [$10\bar{1}1$] (b).

The second one is that TEM images can be obtained in different modes known as bright and dark fields. **Figure 4** shows the image of a deformed sample of a magnesium alloy obtained in the same area, using the same crystal orientation B = [1120] and the same diffraction vector g = [1011]. The difference between both images is that image in **Figure 4(a)** was obtained in dark field mode (image is generated with a diffracted spot) (weak beam), while image in **Figure 4(b)** was obtained in bright field mode (image is generated with the transmitted spot).

Within the magnesium grain, different defects have been created during plastic deformation: dislocations, twins, and subgrains. Twins are clearly observed in bright field mode (**Figure 4b**), and dislocations and subgrains are clearly observed in dark field mode. Therefore, the fractal dimension of the complete deformed substructure changes depending on the TEM mode selected. In general, a combination of images must be used for a complete description of the microstructure. However, there are mechanical tests conducted under specific experimental

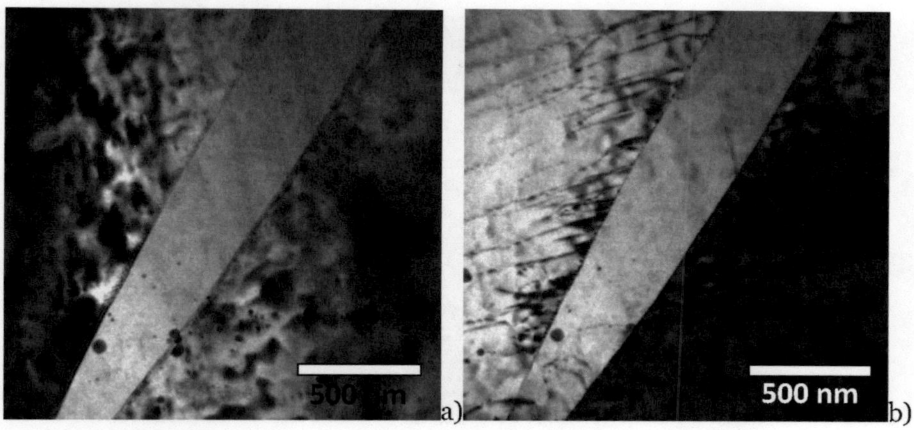

Figure 3.
TEM image of a magnesium alloy deformed at 3% at room temperature and a strain rate of 10^{-3} s^{-1}. Zone axis [$11\bar{2}0$] (a) [0002] and (b) g = [$1\bar{1}01$].

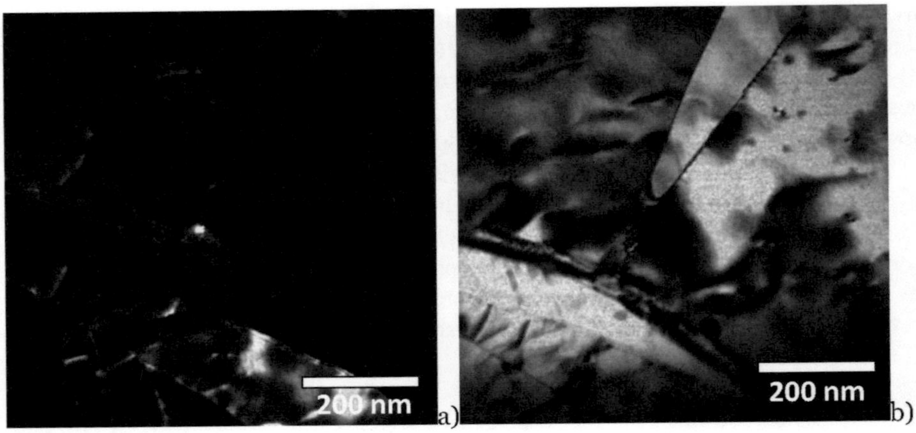

Figure 4.
TEM image of a magnesium alloy deformed at 3% at room temperature and a strain rate of $10^{-3}\ s^{-1}$. Zone axis $[11\bar{2}0]$ g = $[1\bar{1}01]$. (a) Dark field and (b) bright field.

conditions where images from one single mode, bright or dark, are enough for a complete description of the microstructure. In particular, after mechanical testing conducted at high temperature, the dislocation substructure is easily observed in bright field mode because the lattice distortion is almost eliminated.

Finally, it is interesting to point out that TEM images have a 2D nature, whereas dislocations, twins, subgrains, and, in general, the crystal structure have a 3D nature. Cui and Ghoniem [13] studied the influence of size on the fractal dimension of dislocation structure throughout dislocation dynamics simulations. The analysis showed that the D_F of the 3D structure is significantly smaller than the D_F of the 2D corresponding to the projected dislocations in all considered sizes. **Figure 5** shows two images of a 3% deformed magnesium alloy at room temperature of the same area

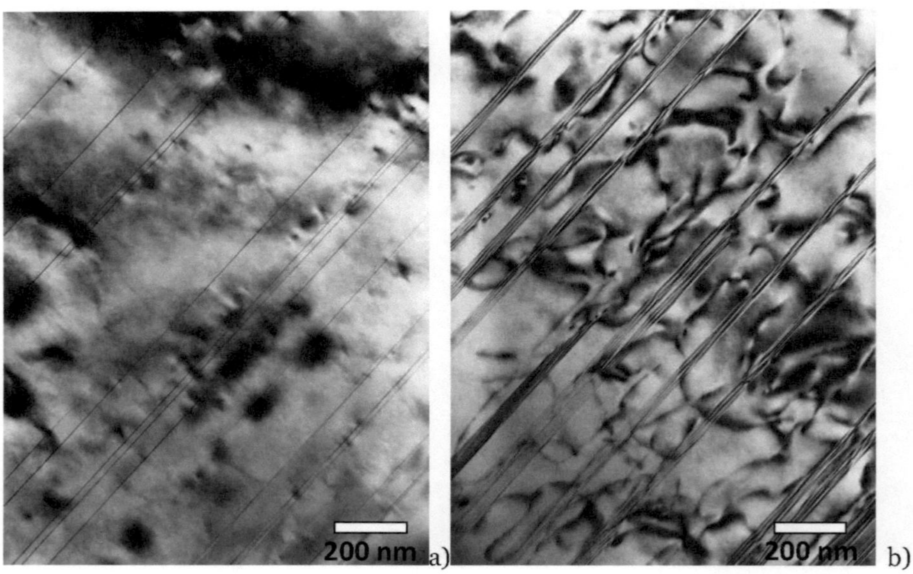

Figure 5.
Bright field image of a magnesium alloy deformed at 3% of plastic strain at room temperature. (a) Sample fully parallel to the beam and (b) sample tilted with respect to the previous condition.

tilted at two different angles. In the first case (**Figure 5a**), a staking fault is observed, and it is fully parallel to the beam. Therefore, atomic lines are clearly observed. When the sample is tilted (**Figure 5b**), it is possible to distinguish the stacking fault throughout the thickness of the TEM sample. Again, the fractal dimension changes.

4. Temperature effect on dislocation fractal structures

The mechanical properties of metals are highly influenced by temperature. In general, the mechanical strength decreases significantly, and the ductility and plastic deformation increases with temperature. In addition to these effects, when $T > 0.5T_{melt}$, time-dependent plasticity becomes dominant [31]. This time-dependent deformation is also called creep. In this temperature range, the dislocations increase their mobility due to the activation of the climbing mechanism, in particular when the staking fault energy is high. The activation of dislocation climbing is revealed in the appearance of well-ordered dislocation structures generated during deformation.

In the following section, some examples of dislocation structures generated at room temperature and high temperature (573 K) will be shown. In the second case, the temperature effect on the arrangement of dislocations to form subgrain boundaries will be observed.

Figure 6.
Bright field image of tangled dislocations in pure aluminum deformed up to 22.3% of plastic strain at room temperature at $10^{4-}s^{-1}$.

4.1 Low-temperature dislocation structures

In **Figure 6**, a bright field TEM image of a commercially pure aluminum single crystal strained up to 22.3% at room temperature at $10^{4-} s^{-1}$ strain rate is shown. This figure corresponds to a perpendicular section of the sample with respect to the tensile axis. Dislocations are arranged in a messy way. The subgrain size developed by deformation of this pure metal at room temperature is around 1 micron.

4.2 High-temperature dislocation structures

In **Figure** 7, a detail of a subgrain boundary shows a bright field TEM image of a commercially pure polycrystalline aluminum strained around 2% at 573 K and 29 MPa. **Figure** 7 corresponds also to a perpendicular section of the sample with respect to the tensile axis. The dislocations are well-ordered in the subgrain boundary. The subgrain size developed by deformation of this pure metal at room 573 K and 29 MPa is around 5 microns.

4.3 Age-hardenable alloys

A very important family of metals is that formed by age-hardenable alloys. Their properties can be tuned by means of heat treatments that usually start with a quenching process of the sample from above the solution temperature. The quenching step is followed by an isothermal annealing that typically lasts for some hours. The variation of mechanical properties is related with the precipitation state developed by the heat treatment. Therefore, deformation that takes place at high temperature is accompanied by an evolution of the precipitation state. The two phenomena are coupled. The deformation slows down due to the presence of atoms in solid solution and coherent precipitates, and the precipitation kinetics is modified

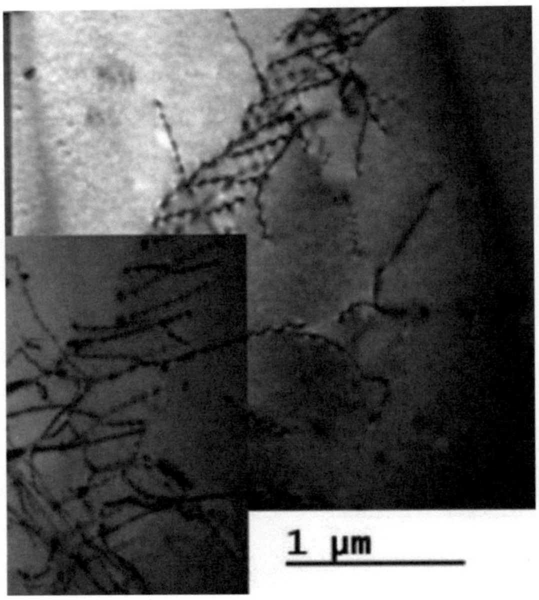

Figure 7.
Bright field image of a subgrain boundary of polycrystalline pure aluminum deformed around 2% at 573 K and 29 MPa.

Figure 8.
Back scattered image of a AA2014 powder particle showing the precipitate structure after quenching from 800 K, followed by aging at 523 K for 10 h.

by the presence of the dislocation structure [32]. **Figure 8** shows the structure of precipitates in aged AA2014.

5. Conclusions

The formation of dislocation fractal structures with plastic deformation is, as in many others aspects of nature, a widespread phenomenon. It is, in fact, observed under different tensing conditions and metal alloys. This is because metal plasticity occurs on several temporal and spatial scales.

On the temporal scale, the dislocations can move at speeds greater than that of the sound when the dislocation density is low, leading to rapid plastic deformation events (e.g., car accident). On the contrary, deformation under creep conditions can occur in much longer time periods, as long as days, months, or even years (e.g., motion of glaciers). The same concept can be applied to the spatial scale as deformation phenomenon involves entities as small as dislocations and vacancies in the crystal lattice, and, on the other side, this phenomenon takes place at a macroscopic scale, meters in size in large components.

The behavior of dislocations during strain of metals is very rich and complex because the plasticity phenomenon covers different spatial and temporal scales and there is a very broad range of experimental conditions and alloys. Nonetheless, the existence of fractal structures of dislocations is ubiquitously found when a metal is deformed. This fact greatly encourages the use of fractals to fully describe the phenomenon of plasticity in metals.

The determination of the fractal dimension of dislocation structures, D_F, is conducted from images obtained by different microstructural characterization methods, typically, by TEM. These images must, then, be carefully treated to minimize inaccuracies in the determination of D_F. The box counting technique is revealed to be a very suitable and reliable method to determine D_F. Although the process is automated in some image analysis applications, the presence of elements different from dislocations (e.g., dots related with other microstructural features) can greatly modify the fractal dimension.

Acknowledgements

This study was financially supported by Proyects MAT2017-83825-C4-1-R from MINECO and Y2018/NMT-4668 from CAM, Spain. The authors would like to thank Antonio Tomás, Dr. Judit Caballero, and Prof. Paloma Adeva from Electron Microscopy Lab of CENIM for their support and help with the microstructure characterization and Edurne Laurín for her help in TEM sample preparation.

Conflict of interest

The authors declare no conflict of interest.

Author details

Ricardo Fernández*, Gaspar González-Doncel and Gerardo Garcés
Department of Physical Metallurgy, National Centre for Metallurgical Research (CENIM) C.S.I.C., Madrid, Spain

*Address all correspondence to: ric@cenim.csic.es

References

[1] Mandelbrot BB. The Fractal Geometry of Nature. San Francisco: Times Books; 1982. ISBN: 0716711869

[2] Zaiser M, Aifantis EC. Materials instabilities and deformation patterning in plasticity. In: Recent Research Developments in Metallurgical and Materials Sciences. Vol. 3. Trivandrum: Research Signpost; 1999. pp. 79-103

[3] Weia X, Xu M, Chen J, Yu C, Chen J, Lu H, et al. Fractal analysis of Mo and Nb effects on grain boundary character and hot cracking behavior for Ni-Cr-Fe alloys. Materials Characterization. 2018; **145**:65-76. DOI: 10.1016/j.matchar.2018.08.024

[4] Agrisuelas J, García-Jareño JJ, Gimenez-Romero D, Negrete F, Vicente F. The fractal dimension as estimator of the fractional content of metal matrix composite materials. Journal of Solid State Electrochemistry. 2009;**13**:1599-1603. DOI: 10.1007/s10008-008-0743-8

[5] Mandelbrot BB, Passoja D, Paullay AJ. Fractal character of fracture surfaces of metal. Nature. 1984;**308**: 721-722. DOI: 10.1038/308721a0

[6] Budaev VP. Application of innovative materials with stochastic nano-andmicrostructure of the surface: control of turbulent flows in plasma and aerodynamics. In: IOP Conference Series: Journal of Physics: Conference Series 1238 2019; 012048. IOP Publishing; 2019

[7] Sevillano GJ, Bouchad E, Kubin LP. Intrinsic size effects in plasticity by dislocation glide. Scripta Metallugica et Materialia. 1991;**25**:355-360. DOI: 10.1016/0956-716X(91)90192-4

[8] Zaiser M, Bai K, Hahner P. Fractal analysis of deformation-induced dislocation patterns. Acta Materialia. 1999;**47**:2463-2476. DOI: 10.1016/S1359-6454(99)00096-8

[9] Usov VV, Shkatulyak NM. Fractal nature of the dislocation structure of low-alloyed steel after controlled rolling. Russian Physics Journal. 2004;**47**: 1139-1146. DOI: 1064-8887/04/4711-1139

[10] Fernández R, González-Doncel G, Garcés G, Bruno G. Fractional Brownian motion of dislocations during creep deformation of metals. Journal of Applied Physics. 2020

[11] Cui Y, Ghoniem N. Influence of size on the fractal dimension of dislocation microstructure. Metals. 2019;**9**:478-487. DOI: 10.3390/met9040478

[12] Eaker CB et al. Oxidation-mediated fingering in liquid metals. Physical Review Letters. 2017;**119**:174502. DOI: 10.1103/PhysRevLett.119.174502

[13] Agarwala A, Shriya Pai S, Shenoy VB. Fractalized metals. arXiv: 1803.01404v1 [cond-mat.dis-nn]. 2018

[14] Weertman J, Weertman JR. Elementary Dislocation Theory. Oxford: Oxford University Press; 1992. p. 213

[15] Turing AM. The chemical basis of morphogenesis. Philosophical Transactions of the Royal Society of London. Series B. Biological Sciences. 1952;**237**:37-72

[16] Fink D, Biersack JP, Chen JT, Städele M, Tjan K, Behar M, et al. Distributions of light ions and foil destruction after irradiation of organic polymers. Journal of Applied Physics. 1985;**58**:668-676. DOI: 10.1063/1.336180

[17] Kurakin A. The self-organizing fractal theory as a universal discovery method: The phenomenon of life. Theoretical Biology & Medical Modelling. 2011;**8**: 4-70. DOI: 10.1186/1742-4682-8-4

[18] de Groot SR, Mazur P. Non-equilibrium thermodynamics. In: Dover Books on Physics. New York, NY; 1985

[19] Goodey NM, Benkovic SJ. Allosteric regulation and catalysis emerge via a common route. Nature Chemical Biology. 2008;**4**:474-482. DOI: 10.1038/nchembio.98

[20] Bogan AA, Thorn KS. Anatomy of hot spots in protein interfaces. Journal of Molecular Biology. 1998;**280**:1-9. DOI: 10.1006/jmbi.1998.1843

[21] Wang Y-J, Ishii A, Shigenobu O. Entropic effect on creep in nanocrystalline metals. Acta Materialia. 2013;**61**:3866-3871. DOI: 10.1016/j.actamat.2013.03.026

[22] Thomson R, Levine LE. Theory of strain percolation in metals. Physical Review Letters. 1998;**81**:3884-3887

[23] Kubin LP, Fressengeas C, Ananthakrishna G. Collective behaviour of dislocations in plasticity. Dislocation in Solids. 2002;**11**:101-192. DOI: 10.1016/S1572-4859(02)80008-0

[24] Ananthachirsnan G. Current theoretical approaches to collective behavior of dislocations. Physics Reports. 2007;**440**:113-259. DOI: 10.1016/j.physrep.2006.10.003

[25] He Z. Integer-dimensional fractals of nonlinear dynamics, control mechanisms, and physical implications. Scientific Reports. 2018;**8**:10324. DOI: 10.1038/s41598-018-28669-3

[26] Kilic K, Abiyev RH. Exploiting the synergy between fractal dimension and lacunarity for improved texture recognition. Signal Processing. 2011;**91**: 2332-2344. DOI: 10.1016/j.sigpro.2011.04.018

[27] Napolitano A, Ungania S, Cannata V. Fractal dimension estimation methods for biomedical images. In: MATLAB – A Fundamental Tool for Scientific Computing and Engineering Applications. Vol. 3. Rijeka: IntechOpen; 2012. pp. 161-178. DOI: 10.5772/48760

[28] Block A, von Bloh W, Schellnhuber HJ. Efficient box-counting determination of generalized fractal dimensions. Physical Review A. 1990; **42**:1869-1874. DOI: 10.1103/PhysRevA.42.1869

[29] Landini G, Murray PI, Misson GP. Local connected fractal dimensions and lacunarity analyses of 60 degrees fluorescein angiograms. Investigative Ophthalmology & Visual Science. 1995; **36**:2749-2755

[30] Available from: https://imagej.nih.gov/ij/plugins/fraclac/FLHelp/Introduction.htm

[31] Fernández R, Bruno G, González-Doncel G. Primary and secondary creep in aluminum alloys as a solid state transformation. Journal of Applied Physiology. 2016;**120**:085101. DOI: 10.1063/1.4961524

[32] Fernández R, Bokuchava G, Toda-Caraballo I, Bruno G, Turchenko V, Gorshkova J, et al. Analysis of the combined strengthening effect of solute atoms and precipitates on creep of aluminum alloys. Advance Engineering Materials. 2020. 1901355. DOI: 10.1002/adem.201901355.